GOVERNMENT–INDUSTRY PARTNERSHIPS

PARTNERING AGAINST TERRORISM

SUMMARY OF A WORKSHOP

CHARLES W. WESSNER

Committee on Government-Industry Partnerships
for the Development of New Technologies

Board on Science, Technology, and Economic Policy

Policy and Global Affairs

NATIONAL RESEARCH COUNCIL
OF THE NATIONAL ACADEMIES

THE NATIONAL ACADEMIES PRESS
Washington, D.C.
www.nap.edu

THE NATIONAL ACADEMIES PRESS   **500 Fifth Street, N.W.**   **Washington, DC 20001**

NOTICE: The project that is the subject of this report was approved by the Governing Board of the National Research Council, whose members are drawn from the councils of the National Academy of Sciences, the National Academy of Engineering, and the Institute of Medicine. The members of the committee responsible for the report were chosen for their special competences and with regard for appropriate balance.

Any opinions, findings, conclusions, or recommendations expressed in this publication are those of the author(s) and do not necessarily reflect the views of the organizations or agencies that provided support for the project.

International Standard Book Number 0-309-09428-3 (Book)
International Standard Book Number 0-309-54616-8 (PDF)
Library of Congress Control Number 2005920142

Limited copies are available from the Policy and Global Affairs Division, National Research Council, 500 Fifth Street, N.W., Washington, DC 20001; 202-334-1529.

Additional copies of this report are available from the National Academies Press, 500 Fifth Street, N.W., Lockbox 285, Washington, DC 20055; (800) 624-6242 or (202) 334-3313 (in the Washington metropolitan area); Internet, http://www.nap.edu.

Copyright 2005 by the National Academy of Sciences. All rights reserved.

Printed in the United States of America

THE NATIONAL ACADEMIES
Advisers to the Nation on Science, Engineering, and Medicine

The **National Academy of Sciences** is a private, nonprofit, self-perpetuating society of distinguished scholars engaged in scientific and engineering research, dedicated to the furtherance of science and technology and to their use for the general welfare. Upon the authority of the charter granted to it by the Congress in 1863, the Academy has a mandate that requires it to advise the federal government on scientific and technical matters. Dr. Bruce M. Alberts is president of the National Academy of Sciences.

The **National Academy of Engineering** was established in 1964, under the charter of the National Academy of Sciences, as a parallel organization of outstanding engineers. It is autonomous in its administration and in the selection of its members, sharing with the National Academy of Sciences the responsibility for advising the federal government. The National Academy of Engineering also sponsors engineering programs aimed at meeting national needs, encourages education and research, and recognizes the superior achievements of engineers. Dr. Wm. A. Wulf is president of the National Academy of Engineering.

The **Institute of Medicine** was established in 1970 by the National Academy of Sciences to secure the services of eminent members of appropriate professions in the examination of policy matters pertaining to the health of the public. The Institute acts under the responsibility given to the National Academy of Sciences by its congressional charter to be an adviser to the federal government and, upon its own initiative, to identify issues of medical care, research, and education. Dr. Harvey V. Fineberg is president of the Institute of Medicine.

The **National Research Council** was organized by the National Academy of Sciences in 1916 to associate the broad community of science and technology with the Academy's purposes of furthering knowledge and advising the federal government. Functioning in accordance with general policies determined by the Academy, the Council has become the principal operating agency of both the National Academy of Sciences and the National Academy of Engineering in providing services to the government, the public, and the scientific and engineering communities. The Council is administered jointly by both Academies and the Institute of Medicine. Dr. Bruce M. Alberts and Dr. Wm. A. Wulf are chair and vice chair, respectively, of the National Research Council.

www.national-academies.org

Steering Committee
for Government-Industry Partnerships
for the Development of New Technologies*

Gordon Moore, *Chair*
Chairman Emeritus, *retired*
Intel Corporation

M. Kathy Behrens
Managing Director of Medical
 Technology
RS Investment Management
and STEP Board

Michael Borrus
Managing Director
The Petkevich Group, LLC

Iain M. Cockburn
Professor of Finance and Economics
Boston University

Kenneth Flamm
Dean Rusk Chair in International
 Affairs
LBJ School of Public Affairs
University of Texas at Austin

James F. Gibbons
Professor of Engineering
Stanford University

W. Clark McFadden
Partner
Dewey Ballantine

Burton J. McMurtry
General Partner
Technology Venture Investors

William J. Spencer, *Vice-Chair*
Chairman Emeritus, *retired*
International SEMATECH
and STEP Board

Mark B. Myers
Visiting Professor of Management
The Wharton School
University of Pennsylvania
and STEP Board

Richard Nelson
George Blumenthal Professor of
 International and Public Affairs
Columbia University

Edward E. Penhoet
Chief Program Officer
Science and Higher Education
Gordon and Betty Moore Foundation
and STEP Board

Charles Trimble
Chairman
U.S. GPS Industry Council

John P. Walker
Chairman and Chief Executive
 Officer
Axys Pharmaceuticals, Inc.

Patrick Windham
President, Windham Consulting
and Lecturer, Stanford University

---

*As of October 2002.

Project Staff*

Charles W. Wessner
Study Director

Alan Anderson
Consultant

Christopher S. Hayter
Program Associate

Tabitha M. Benney
Program Associate

Adam Korobow
Program Officer

McAlister T. Clabaugh
Program Associate

Sujai J. Shivakumar
Program Officer

David E. Dierksheide
Program Associate

---

*As of October 2002.

For the National Research Council (NRC), this project was overseen by the Board on Science, Technology, and Economic Policy (STEP), a standing board of the NRC established by the National Academies of Sciences and Engineering and the Institute of Medicine in 1991. The mandate of the STEP Board is to integrate understanding of scientific, technological, and economic elements in the formulation of national policies to promote the economic well-being of the United States. A distinctive characteristic of STEP's approach is its frequent interactions with public and private-sector decision makers. STEP bridges the disciplines of business management, engineering, economics, and the social sciences to bring diverse expertise to bear on pressing public policy questions. The members of the STEP Board* and the NRC staff are listed below.

Dale Jorgenson, *Chair*
Frederic Eaton Abbe Professor of
   Economics
Harvard University

M. Kathy Behrens
Managing Director of Medical
   Technology
RS Investment Management

Bronwyn Hall
Professor of Economics
University of California at Berkeley

James Heckman
Henry Schultz Distinguished Service
   Professor of Economics
University of Chicago

Ralph Landau
Consulting Professor of Economics
Stanford University

Richard Levin
President
Yale University

William J. Spencer, *Vice-Chair*
Chairman Emeritus, *retired*
International SEMATECH

David T. Morgenthaler
Founding Partner
Morgenthaler

Mark B. Myers
Visiting Professor of Management
The Wharton School
University of Pennsylvania

Roger Noll
Morris M. Doyle Centennial Professor
   of Economics
Stanford University

Edward E. Penhoet
Chief Program Officer
Science and Higher Education
Gordon and Betty Moore Foundation

William Raduchel
Chief Technology Officer
AOL Time Warner

Alan Wm. Wolff
Managing Partner
Dewey Ballantine

---

*As of October 2002.

STEP Staff*

Stephen A. Merrill
Executive Director

Russell Moy
Senior Program Officer

Craig M. Schultz
Research Associate

Camille M. Collett
Program Associate

Christopher S. Hayter
Program Associate

David E. Dierksheide
Program Associate

Charles W. Wessner
Program Director

Sujai J. Shivakumar
Program Officer

Adam Korobow
Program Officer

McAlister T. Clabaugh
Program Associate

Tabitha M. Benney
Program Associate

---

*As of October 2002.

**National Research Council
Board on Science, Technology, and Economic Policy**

Sponsors

The National Research Council gratefully acknowledges the support of the following sponsors:

National Aeronautics and Space Administration

Office of the Director, Defense Research & Engineering

National Science Foundation

U.S. Department of Energy

Optoelectronics Industry Development Association

Office of Naval Research

National Institutes of Health

National Institute of Standards and Technology

Sandia National Laboratories

Electric Power Research Institute

International Business Machines

Kulicke and Soffa Industries

Merck and Company

Milliken Industries

Motorola

Nortel

Procter and Gamble

Silicon Valley Group, Incorporated

Advanced Micro Devices

Any opinions, findings, conclusions, or recommendations expressed in this publication are those of the authors and do not necessarily reflect the views of the project sponsors.

Reports in the Series

**Government-Industry Partnerships
for the Development of New Technologies**

New Vistas in Transatlantic Science and Technology Cooperation
Washington, D.C.: National Academy Press, 1999

*Industry-Laboratory Partnerships: A Review of the Sandia Science and
Technology Park Initiative*
Washington, D.C.: National Academy Press, 1999

The Advanced Technology Program: Challenges and Opportunities
Washington, D.C.: National Academy Press, 1999

*The Small Business Innovation Research Program:
Challenges and Opportunities*
Washington, D.C.: National Academy Press, 1999

*The Small Business Innovation Research Program: An Assessment of the
Department of Defense Fast Track Initiative*
Washington, D.C.: National Academy Press, 2000

A Review of the New Initiatives at the NASA Ames Research Center
Washington, D.C.: National Academy Press, 2001

The Advanced Technology Program: Assessing Outcomes
Washington, D.C.: National Academy Press, 2001

*Capitalizing on New Needs and New Opportunities: Government-Industry
Partnerships in Biotechnology and Information Technologies*
Washington, D.C.: National Academy Press, 2002

Partnerships for Solid-State Lighting
Washington, D.C.: National Academy Press, 2002

*Government-Industry Partnerships for the Development of New Technologies:
Summary Report*
Washington, D.C.: National Academies Press, 2002

*Securing the Future: Regional and National Programs to Support the
Semiconductor Industry*
Washington, D.C.: National Academies Press, 2003

Contents

PREFACE		xv
I.	**INTRODUCTION**	1
II.	**PROCEEDINGS**	21

Welcome 23
Bruce Alberts, National Academy of Sciences

Introduction 25
William Spencer, International SEMATECH

Panel I: Partnering to Meet the New Security Challenge 28
  *Moderator: Sean O'Keefe, National Aeronautics and
    Space Administration*

  **Partnering for Cyber Security and Infrastructure
    Protection** 29
  Congressman Sherwood L. Boehlert (R-NY)

  Capitalizing on the Nation's Research Portfolio 34
  Gordon Moore, Intel Corporation

xi

Panel II: Best Practice Examples of Public-Private Partnerships   38
Moderator: Arden Bement, National Institute of Standards and Technology

SEMATECH: Assessing the Contribution   41
Kenneth Flamm, University of Texas at Austin

Partnering for Progress: The Advanced Technology Program   48
Maryann Feldman, Johns Hopkins University

University Research and the Market: The Carnegie Mellon Experience   51
Christina Gabriel, Carnegie Mellon University

Discussant: Michael Borrus, The Petkevich Group, LLC   60

Panel III: Partnerships Against Bioterrorism   66
Moderator: Larry Kerr, Department of Homeland Security

Partnering for Vaccines: The NIAID Perspective   67
Carol Heilman, National Institute of Allergy and Infectious Diseases

Partnering for Counter Measures: The Private Research Perspective   75
Gail Cassell, Lilly Research Laboratories, Eli Lilly & Company

Discussant: Kathy Behrens, RS Investment Management   80

Panel IV: Partnering for National Security   87
Moderator: William B. Bonvillian, Office of Senator Lieberman

Overcoming Information Overload   91
Anne K. Altman, IBM Corporation

New Technologies for New Threats   95
Ronald M. Sega, Department of Defense

Security Challenges in an Open Economy   101
Steve Flynn, Council on Foreign Relations

**Panel V: Roundtable on Partnering for National Missions:
Defense, Health and Energy** 109
Moderator: Patrick Windham, Windham Consulting

Christina Gabriel, Carnegie Mellon University
William Spencer, International SEMATECH
William Bonvillian, Office of Senator Lieberman
James Turner, House Science Committee

Closing Remarks 114
Gordon Moore, Intel Corporation

III. APPENDIXES

A. Biographies of Speakers 117

B. Participants List 132

C. Bibliography 138

Preface

The National Academies has sought to bring the nation's great strength in science and technology to bear on protecting the United States against terrorism. In a major 2002 report, *Making the Nation Safer: The Role of Science and Technology in Countering Terrorism*, the National Academies characterized the range of threats to the nation's security and identified research agendas to strengthen areas of vulnerability. It also outlined policies needed to strengthen the government's ability to draw on the nation's capacities in science and technology for combating terrorism. Specifically, it noted that effective public-private partnerships must occur for the government and private sector to work together to enhance homeland security.[1]

In recent years public-private partnerships have played an increased role in developing new technologies both in the United States and abroad. To further our understanding of the motivations, operations, and policy challenges associated with public-private partnerships, the National Research Council's Board on Science, Technology, and Economic Policy (STEP) launched in 1998 a major review of U.S. and foreign programs. This program-based analysis was led by Gordon Moore, Chairman Emeritus of Intel, and Bill Spencer, Chairman Emeritus of International SEMATECH. It was carried out by a distinguished multidisciplinary Steering Committee that included members from academia, high-technology industries, venture capital firms, and the realm of public policy. The Committee's analysis—which included a significant (though necessarily limited) portion of the variety of cooperative activity that takes place between the government and

---

[1] See National Research Council, *Making the Nation Safer: The Role of Science and Technology in Countering Terrorism*, Lewis M. Branscomb and Richard D. Klausner, eds., Washington, D.C.: The National Academies Press, 2002.

the private sector—focused on "best practices" among major U.S. partnerships as a way of drawing out positive guidance for future public policy.[2]

At its concluding conference on October 2, 2002, the National Research Council Committee on Government-Industry Partnerships drew together the findings of its four-year study on partnerships to explore how public-private partnerships can help make the nation safer against terrorism. The conference was well received. Subsequently, the Governing Board Executive Committee authorized the release of a summary report of the workshop. Accordingly, this report summarizes the proceedings of that conference, along with an introductory chapter that highlights key issues raised at the conference. These issues are central to the country's ongoing efforts to develop new technologies and new approaches to meet the terrorist threat.

ACKNOWLEDGMENTS

On behalf of the National Academies, we express our appreciation and recognition for the insights, experiences, and perspectives of the conference participants. A number of individuals deserve recognition for their contributions to the preparation of this report. Foremost among these were Dr. Sujai Shivakumar and Alan Anderson, who played an instrumental role in the preparation of this report. Others to whom recognition is due include Christopher Hayter, David Dierksheide, Tabitha Benney, and McAlister Clabaugh. Without their collective efforts, amidst many other competing priorities, it would not have been possible to prepare this report.

NATIONAL RESEARCH COUNCIL REVIEW

This report has been reviewed in draft form by individuals chosen for their diverse perspectives and technical expertise, in accordance with procedures approved by the NRC's Report Review Committee. The purpose of this independent review is to provide candid and critical comments that will assist the institution in making its published report as sound as possible and to ensure that the report meets institutional standards for quality and objectivity. The review comments and draft manuscript remain confidential to protect the integrity of the process. We wish to thank the following individuals for their review of this report: F. M. Ross Armbrecht, Jr., President of the Industrial Research Institute, Howard Frank, Dean of the Robert H. Smith School of Business at the University

---

[2]The findings and recommendations of the Committee's study on public-private partnerships are presented in National Research Council, *Government-Industry Partnerships for the Development of New Technologies: Summary Report*, C. Wessner, ed., Washington, D.C.: The National Academies Press, 2003.

of Maryland, Lewis S. Edelheit, Retired Senior Research & Technology Advisor, General Electric Company, and Christina Gabriel, Vice Provost and Chief Technology Officer, Carnegie Mellon University. Although the reviewers listed above have provided many constructive comments and suggestions, they were not asked to endorse the content of the report, nor did they see the final draft before its release. The review of this report was overseen by John White of the John F. Kennedy School of Government, Harvard University, who was responsible for making certain that an independent examination of this report was carried out in accordance with institutional procedures and that all review comments were carefully considered. Responsibility for the final content of this report rests entirely with the author and the institution.

Charles W. Wessner

I
INTRODUCTION

Introduction

"For the government and private sector to work together on increasing homeland security, effective public-private partnerships and cooperative projects must occur."

*Making the Nation Safer: The Role of Science
and Technology in Countering Terrorism*
National Research Council, 2002

PUBLIC-PRIVATE PARTNERSHIPS IN THE WAR ON TERROR

The National Academies' response to the threat of terrorism has been to bring the nation's great strength in science and technology to bear on protecting the United States.[1] In its June 2002 report, *Making the Nation Safer: The Role of Science and Technology in Countering Terrorism*, the National Academies recommended that effective public-private partnerships must occur for the government and private sector to work together on increasing homeland security.[2] Following on this recommendation, the National Academies' Committee on Government-Industry Partnerships for the Development of New Technologies, led by Gordon Moore, drew together the findings of its four-year study at its final conference to explore how partnerships can contribute to the nation's present war on terror. This chapter introduces the main points of that conference. The conference proceedings are summarized in the next chapter.

Partnerships are cooperative relationships involving government, industry, laboratories, and (increasingly) universities organized to encourage innovation and commercialization. Partnerships come in many forms, including industry con-

---

[1] See opening remarks by Dr. Bruce Alberts, President of The National Academies, in the Proceedings chapter of this volume.

[2] See National Research Council, *Making the Nation Safer: The Role of Science and Technology in Countering Terrorism*, Lewis M. Branscomb and Richard D. Klausner, eds., Washington, D.C.: The National Academies Press, 2002.

Box A: Why Partnerships are Crucial Now

1) *The challenge of responding to the threat of terrorism is unique.* As Congressman Boehlert noted in his conference presentation, the government must provide security, and there is as yet no market for many of the products or services required. To meet this "market failure," he emphasized the need for partnerships among industry, universities, and national laboratories to develop solutions to unique challenges of homeland security.
2) *The need for speed.* Following 9/11, the nation resolved that every effort should be made to keep such terrorist attacks from recurring. This requires that solutions to the varied challenges of homeland security be developed as rapidly as possible. Partnerships are a way of rapidly mobilizing the knowledge base of the nation and focusing it on new national needs. Carol Heilman of NIH noted in her conference presentation that the Small Business Innovation Research (SBIR) program, a federal public-private partnership, has proven to be effective in quickly mustering the expertise dispersed across the country to address specific national security needs.
3) *The need for products.* As the NRC report *Making the Nation Safer* notes, partnerships such as SBIR and the Advanced Technology Program (ATP) can focus on the development of concrete products that can be deployed in the war on terror.[a] In addition, the Defense Advanced Research Projects Agency (DARPA) was cited by Congressman Boehlert and by Steven Kerr of the Department of Homeland Security as an organizational model for stimulating new thinking and applied research that is focused on new products. Dr. Kerr noted that the Homeland Security version of DARPA—HSARPA—would be a "major facilitator to couple the research and development testing and evaluation enterprise with the actual entities, whether they be in the private sector or in academia, and the actual end-users."
4) *Using proven mechanisms.* Each of these needs can be met through the use of existing proven partnerships, such as SBIR, ATP, and DARPA. Their established procedures and proven track record make them well suited to meeting the new challenges of the war on terror.

---

[a]See National Research Council, *Making the Nation Safer: The Role of Science and Technology in Countering Terrorism*, Lewis M. Branscomb and Richard D. Klausner, eds., Washington, D.C.: The National Academies Press, 2002.

sortia, innovation awards, and university and laboratory-based science and technology clusters. The Committee found that such public-private partnerships, when properly crafted, can help usher in the development of new processes, products, and services. With well-managed partnerships, the government can realize missions in health, environmental protection, and national security, often leveraging lower cost or more effective technologies.[3]

Appropriately structured partnerships can also serve as a policy instrument that aligns the incentives of private firms to achieve national missions without compelling them to do so. As the 2002 National Academies report on countering terrorism notes, "A more effective approach is to give the private sector the widest possible latitude for innovation and, where appropriate, to design R&D strategies in which commercial uses of technologies rest on a common base of investment. Companies then have the potential to address vulnerabilities while increasing the robustness of public and private infrastructure against unintended and natural failures, improving the reliability of systems and quality of service, and in some cases, increasing productivity."[4]

A Tested Policy Tool to Enhance National Security

This cooperative public-private approach is not new, as Dr. Moore concluded in his conference presentation, noting that partnerships have "helped us to meet some of our national missions; they've been used effectively since the founding of our country; and our studies have identified some useful features in producing results. The appropriate thing now is to apply what we've learned to meet the challenges posed by terrorism." The examples below illustrate some key technologies that helped meet critical national missions that were fostered through public-private partnerships.

As the Committee's Summary Report of its ten-volume study points out, public-private partnerships have long been employed as a policy tool to help protect U.S. national security. These investments have led, moreover, to major new industries, often contributing significantly to the nation's economic prosperity. An illustrative list of technologies advanced through various partnership mechanisms includes:

- Muskets: The federal government's contract for the unprecedented concept of interchangeable musket parts was made to the inventor Eli Whitney in 1798. The ultimate success of this approach laid the foundation of the first ma-

---

[3]See National Research Council, *Government-Industry Partnerships for the Development of New Technologies: Summary Report*, C. Wessner, ed., Washington, D.C.: The National Academies Press, 2003.

[4]See National Research Council, *Making the Nation Safer: The Role of Science and Technology in Countering Terrorism*, op. cit., p. 360.

chine tool industry,[5] meeting both U.S. defense needs and contributing to American industrialization.

- Telegraph: A similar award of $30,000 made by the Congress in 1842 enabled Samuel Morse to demonstrate the feasibility of the telegraph. His success transformed communications for military and civil needs in the decades that followed.[6]
- Railroads: The federal government played an instrumental role in developing the U.S. railway network through the Pacific Railroad Act of 1862 and the Union Pacific Act of 1864.[7] Private enterprise lacked the means to construct transcontinental railroads without the substantial federal support provided by these Acts. The railways transformed the American economy while also integrating the western territories with the East.

---

[5]The 1798 contract with Eli Whitney was an early example of high-technology procurement. Whitney missed his first delivery date for the arms and encountered substantial cost overruns, a set of events that is still familiar. However, his focus on the concept of interchangeable parts and the machine tools to make them was prescient. David A. Hounshell in his excellent analysis of the development of manufacturing technology in the United States suggests that Simeon North was in fact the one who succeeded in achieving interchangeability and the production of components by special-purpose machinery. See *From the American System to Mass Production 1800–1932*, Baltimore: Johns Hopkins University Press, 1985, pp. 25–32. By the 1850s, the United States had begun to export specialized machine tools to the Enfield Arsenal in Great Britain. The British described the large-scale production of firearms, made with interchangeable parts, as "the American system of manufacturers." See David C. Mowery and Nathan Rosenberg, *Paths of Innovation: Technological Change in 20th Century America*, New York: Cambridge University Press, 1998, p. 6. Whitney's concept of interchangeable parts, and the machine tools to make them was in the end successful.

[6]For a discussion of Samuel Morse's 1837 application for a grant and the congressional debate, see Irwin Lebow, *Information Highways and Byways*. New York: Institute of Electrical and Electronics Engineers, 1995, pp. 9–12. For a more detailed account, see Robert Luther Thompson, *Wiring a Continent: The History of the Telegraph Industry in the United States 1823–1836*. Princeton, N.J.: Princeton University Press, 1947.

[7]For an economic history of the transcontinental railroad, see Robert W. Fogel, *Railroads and American Economic Growth: Essays in Econometric History*, Baltimore: Johns Hopkins University Press, 1964. See also Alfred P. Chandler, *Strategy and Structure: Chapters in History of the Industrial Enterprise*, Cambridge, MA: MIT Press, 1962. For a popular historical account, see Stephen Ambrose, *Nothing Like It in the World: The Men Who Built the Transcontinental Railroad 1863–1869*, New York: Simon and Schuster, 2000. In the midst of the Civil War, Abraham Lincoln signed the Pacific Railroad Act of 1862 providing the necessary standards and substantial incentives to launch the first transcontinental railroad. Financial aid to the railroads was provided in the form of government bonds at $16,000 to $48,000 per mile depending on terrain, as well as land grants for stations, machine shops, etc. In addition, right of way was to extend 200 feet on both sides of the road. The Pacific Railroad Act was supplemented in 1864 by the Union Pacific Act, which did not increase government funding but allowed the railroad companies to issue their own first-mortgage bonds. This act also allowed President Lincoln to set the "standard gauge" at 4 feet, 8 1/2 inches. As with fiber-optic investments today, there was some overbuilding, but the fundamental policy objectives of national unity and economic growth were achieved. From 30,000 miles of railway in 1860 rail mileage grew to more than 201,000 by 1900, linking the nation together.

- Aircraft: In 1903, the Wright Brothers, meeting the terms of an Army contract, demonstrated the feasibility of manned flight. Later, the National Advisory Committee for Aeronautics, formed in 1915, made major contributions to the development of the U.S. civil and military aircraft industry.[8]
- Radio manufacturing: RCA, founded in 1919 on the initiative of the U.S. Navy, served both commercial and military needs for a U.S.-based radio industry.[9]
- Computers: During the Second World War, U.S. investments resulted in the creation of the ENIAC, one of the earliest digital computers. In the postwar period, military investments and encouragement played an instrumental role in developing the fledgling American computer industry.[10] Military needs and the private sector's ability to meet them provided the foundation for the growth of the information economy.
- Internet: Government investments, both civil and military, were crucial to the development of today's Internet. These investments were made over a sustained period of time, in close cooperation with leading university researchers and the private sector, with the ultimate applications not clearly foreseen. While the military, economic, and social transformations resulting from these investments are still unfolding, the Internet demonstrated an immediate benefit during the attacks of 11 September 2001 by providing resilience and redundancy to the U.S. communication system.[11]

---

[8] See D. Mowery and N. Rosenberg, *Technology and the Pursuit of Economic Growth*, New York: Cambridge University Press, 1989, Chapter 7, especially pp. 181–194. The authors note that the commercial aircraft industry is unique among manufacturing industries in that a federal research organization, the National Advisory Committee on Aeronautics (NACA, founded in 1915 and absorbed by NASA in 1958), conducted and funded research on airframe and propulsion technologies. Before World War II, NACA operated primarily as a test center for civilian and military users. NACA made a series of remarkable contributions regarding engine nacelle locations and the NACA cowl for radial air-cooled engines. These innovations, together with improvements in engine fillets based on discoveries at Caltech and the development of monocoque construction, had a revolutionary effect on commercial and military aviation. These inventions made the long-range bomber possible, forced the development of high-speed fighter aircraft, and vastly increased the appeal of commercial aviation. See Lebow, *Information Highways and Byways*, op. cit.; and Alexander Flax, National Academy of Engineering, personal communication, September 1999. See also Roger E. Bilstein, *A History of the NACA and NASA, 1915–1990*, Washington, D.C.: National Aeronautics and Space Administration Office of Management Scientific and Technical Information Division, 1989.

[9] Josephus Daniels, Secretary of the Navy during the Wilson Administration, appeared to feel that monopoly was inherent to the wireless industry, and if that were the case, he believed the monopoly should be American. By pooling patents, providing equity, and encouraging General Electric's participation, the Navy helped to create the Radio Corporation of America. See Irwin Lebow, *Information Highways and Byways*, pp. 97–98 and Chapter 12. See also Michael Borrus and Jay Stowsky, "Technology Policy and Economic Growth," BRIE Working Paper 97, April 1997.

[10] See Kenneth Flamm, *Creating the Computer*, Washington, D.C.: Brookings, 1988.

[11] For an excellent review of the role of government support in developing the computer industry and the Internet, see National Research Council, *Funding a Revolution: Government Support for Computing Research*, Washington, D.C.: National Academy Press, 1999.

Indeed, as Dr. Moore observed in his conference presentation, partnerships are essential if the nation is to capitalize on its research portfolio in addressing the problem of terrorism. The United States, he said, had "the best and broadest science and technology in the world," noting that the task ahead lay in applying this knowledge to the challenge of securing the nation against the threat of terrorism. He cited several areas where sufficient technical knowledge may be developed through public-private partnerships, including the development of new instrumentation to detect radiation at large distances and the rapid identification of bio-agents including vaccines, antibiotics, and anti-viral agents. "The solution," he concluded, "requires partnerships—it requires the best minds, including the flexibility to include future technology; it needs an adequate budget, built with off-the-shelf software and hardware products; and it probably requires a short-term and long-term strategy that will be most effective in handling the security problems we have to anticipate."

Some Characteristics of Successful Partnerships

How can we apply what we have learned about partnerships to meet the challenge posed by terrorism? The NRC Committee found that "properly constructed, operated, and evaluated partnerships can provide an effective means for accelerating the progress of technology from the laboratory to the market."[12] At the conference, Bill Spencer and Michael Borrus highlighted many of the specific features that the Committee found to be necessary for a successful partnership.[13]

- **Clear and measurable set of objectives:** Bill Spencer noted that a partnership's objectives should be established, measured, and reported on regularly. Referring to the experience of the Sematech semiconductor consortium, he noted that objectives should be focused as closely as possible on generic or pre-competitive work, rather than on products closer to the commercial market.
- **Frequent, rigorous evaluation:** Michael Borrus pointed out that regular, external, and objective assessments, with a willingness to stop failed projects, are necessary for a partnership to succeed.
- **Flexibility:** Michael Borrus also noted that a willingness to adjust to new technologies and new market opportunities is necessary for a partnership's suc-

---

[12]See National Research Council, *Government-Industry Partnerships for the Development of New Technologies: Summary Report,* op. cit., Finding VII, p. 29.

[13]For additional discussion, see National Research Council, *Government-Industry Partnerships for the Development of New Technologies: Summary Report,* op. cit., "Conditions for Successful Partnerships," pp. 13–16.

cess. Indeed, this flexibility, according to Dr. Spencer, was instrumental in Sematech's success.[14]

- **Quality industry-initiated leadership:** Industry leadership and cost-sharing helps ensure that the industry partner is an active participant and has a stake in a positive outcome. Industry leadership provides the partnership with the technical expertise, experienced management, flexibility, and credibility needed for success. As Bill Spencer emphasized, partnerships need to be led by the "very best people in the industry involved."

- **Adequate funding:** As Michael Borrus observed, those projects that succeed tend to have funding commensurate with their goals; either too much or too little funding can impede progress.

NEW THREATS AND NEW RESPONSES

The role of partnerships in securing the nation against a variety of threats, ranging from economic terrorism and cyber-terrorism to bioterrorism was discussed at the conference.

Securing Ports and International Commerce

Steve Flynn of the Council on Foreign Relations described the vulnerabilities of the international container shipment system and discussed a possible partnership that could make international shipping both more secure and more efficient. Although container shipping is a cornerstone of today's global economy, security is not built into this transportation system. Mr. Flynn noted that there are some 16 million cargo containers in use around the world, which are easy to purchase, fill with cargo, and—with minimum documentation—deliver to any container port in the world. An effort to stop and inspect all such containers in the United States would take about six months and effectively tie up global commerce—with potentially a much larger negative economic fallout than any particular terrorist strike itself.

An alternative approach is to implement a new system that reliably reports on the integrity of a given container and that tracks its movements to make sure that it has not been intercepted or tampered with. Such a system, which has been demonstrated to be technically feasible, could both improve security as well as make the world of supply chains and international logistics more efficient. Mr. Flynn advocated a public-private partnership mechanism that integrates available

---

[14]Sematech is the semiconductor industry consortium, widely regarded as having contributed to the resurgence of the U.S. semiconductor industry in the decade of the 1990s. See National Research Council, *Securing the Future: Regional and National Programs to Support the Semiconductor Industry*, C. Wessner, ed., Washington, D.C.: The National Academies Press, 2003.

technologies (e.g., satellite tracking) with operational realities best known to employees of port authorities, U.S. attorneys, and others working in the field. With support from Washington, such a partnership can address the urgent government mission of physically securing the nation while also safeguarding its economic foundations.

Enhancing Cyber Security

As with the case of container shipment, the software market has favored speed, ease of use, and low prices over security, according to Representative Sherwood Boehlert. This had led to inadequate technical knowledge about designing secure computers and computer networks, and a lack of wherewithal to proceed. This attitude has now begun to change, he noted, as cyber-security has become a hot topic on Capitol Hill, and elsewhere.

Mr. Boehlert noted that creative partnerships are needed to foster new ideas for cyber security. The government has to be involved, he said, because improving cyber security requires more basic research, and will require greater support for students in order to attract new people to the computer security field. Academia has to be involved because much of the expertise in this area resides in colleges and universities, which have the capacity to educate a cadre of computer security experts. Finally, industry has to be involved because private firms bring an essential perspective as to what is needed in this rapidly evolving field, and because advances in computer security must be able to succeed in the private marketplace if they are to have the desired impact. Congressman Boehlert predicted that new legislation—the Computer Security Research and Development Act—would create, to this end, new partnership programs at the National Science Foundation and the National Institute of Standards and Technology.

Preparing for Bioterrorism

Carole Heilman of the National Institutes of Health (NIH) and Gail Cassell of Eli Lilly addressed the unique challenges the nation faces in preparing for possible terrorist attacks that use biological agents. Dr. Heilman pointed out that the challenge is one of preparing for a threat that can take a range of forms, that affects a heterogeneous population in multiple ways, and that requires a rapid response. Similarly, Dr. Cassell noted the potential diversity of biological weapons, including a number of different viruses and bacterial agents, and many infectious agents. Responding to a threat of this complexity, she said, requires being prepared to develop and administer a broad-spectrum of therapies.

Developing effective vaccines is a long and tedious process, noted Dr. Heilman, because vaccines have difficult biologics and because regulatory hurdles slow the process of research. For NIH to develop an adequate portfolio and stock of vaccines for bio-defense, she said, public-private partnerships between NIH

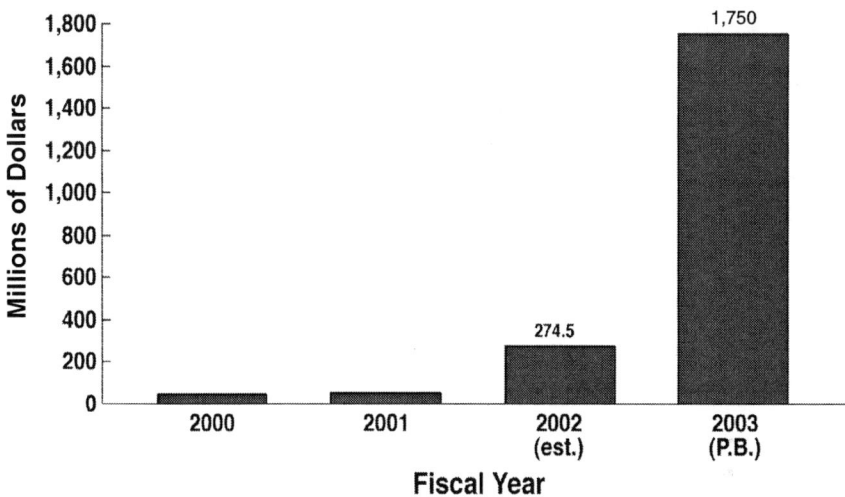

FIGURE 1 NIH biodefense research funding, FY 2000–2003.

and the private sector are both technically and financially necessary. Preparing an adequate bio-defense also requires the development of new anti-viral treatments and antibiotics, added Dr. Cassell. At present, health officials have at their disposal only a small number of anti-virals to naturally occurring viruses. The situation for antibiotics, is marginally better, she noted, but still worrisome given that only two new classes of antibiotics have been developed and introduced over the past thirty to forty years. Yet, private firms attempting to develop new anti-virals and antibiotics face not only daunting technical challenges, but also serious financial hurdles—especially given that the failure rate for drug discovery averages at about ninety percent, with failure rates for antibiotic drug discovery ranging even higher. Recognizing that these hurdles dampen private sector enthusiasm for drug discovery, noted Dr. Cassell, partnerships are needed to share the high financial risks and to pool dispersed knowledge about health risks.

Addressing the complex challenge of biodefense requires a major financial commitment. Dr. Heilman noted that Congress understands that a sustained high level of support is needed since homeland biodefense can only be realized through a long-term commitment of support. She also presented a graph (Figure 1) showing a six-fold increase in NIH's biodefense budget.

PARTNERING MECHANISMS TO MEET NEW SECURITY CHALLENGES

For the current war on terrorism, partnerships have a demonstrated capacity to marshal the ingenuity of industry to meet new needs for national security.

Because they are flexible and can be organized on an *ad hoc* basis, partnerships can be an effective means to focus diverse expertise and innovative technologies to help counter new threats. Indeed, the National Academies 2002 report *Making the Nation Safer* identified several existing models for government-industry collaboration that could contribute to the war on terror, including the Advanced Technology Program (ATP) and the Small Business Research Innovation (SBIR) program.[15] Existing programs with established selection procedures and mechanisms for granting and evaluation of awards offer major benefits in comparison to founding completely new programs; notably, they can "hit the ground running."

Reflecting this reality, the roles that ATP and SBIR might play in developing technologies to counter terrorism were examined at the conference. In addition, the experience of the Sematech semiconductor consortium in providing best practice lessons for effective partnerships and in securing the nation's capability in a key technology were reviewed.

The Advanced Technology Program

As Maryann Feldman of Johns Hopkins University noted, ATP was initiated as a means of funding high-risk R&D with broad commercial and societal benefits that would not be undertaken by a single company, either because the risk was too high or because a large enough share of the benefits of success would not accrue to the company for it to make the investment.

Specifically, ATP provides cost-shared funding to industry intended to accelerate the development and dissemination of high-risk technologies with the potential for broad-based economic benefits for the U.S. economy.[16] ATP funding is directed to technical research (but not product development). Companies, whether singly or jointly, conceive, propose, and execute all projects, often in collaboration with universities and federal laboratories. The ATP shares the project costs for a limited time. Single-company awardees can receive up to $2 million for R&D activities for up to three years. Larger companies must contribute at least 60 percent of the total project cost. Joint ventures can receive funds for R&D activities for up to five years.[17] Since 1992, ATP has obligated an estimated $270 million to companies and joint ventures pursuing promising commercial ventures that could be enlisted in the fight against terrorism, according to Arden

---

[15]See National Research Council, *Making the Nation Safer: The Role of Science and Technology in Countering Terrorism*, op. cit., 2002, p. 359.

[16]See National Research Council, *The Advanced Technology Program: Challenges and Opportunities*, C. Wessner, ed., Washington, D.C.: National Academy Press, 1999.

[17]For a discussion of the ATP program and its role in the U.S. innovation system, and an evaluation of its contributions, see National Research Council, *The Advanced Technology Program: Assessing Outcomes*, C. Wessner, ed., Washington D.C.: National Academy Press, 2001.

Bement, of the National Institute of Standards and Technology. These technologies, he said, underscore the dual nature of many of the technologies that are now needed.

Dr. Feldman suggested that because it is industry driven, with the ideas and half the funds coming from the private sector, ATP can have a positive impact on developing new technologies for the war on terror. Drawing on her empirical analyses of the program, she noted that ATP selects those projects that had greater potential to generate substantial public benefits—primarily riskier, early-stage projects and new partnerships. In addition, Dr. Feldman found that ATP increases private sector R&D in the kinds of activities it funds. Garnering an ATP award bestows a "halo effect" that makes it easier for participating firms to raise subsequent funding. In effect, an ATP award creates additional information for private investors in the risky market of early-stage finance. A further advantage is that ATP relies on commercial firms to propose projects. This bottom-up approach encourages the private sector to identify and invest in the kinds of applications that have the highest potential to bring new capabilities to bear on national security concerns.[18]

The Small Business Innovation Research Program

This early-stage technology development program can be a useful mechanism to draw ideas from small companies and university research and connect them to market needs and agency missions. Noting that the university community has shown strong interest in adapting their research to serve the needs of the nation in the wake of the September 11 attacks, Christina Gabriel of Carnegie Mellon University outlined initiatives that Carnegie Mellon was taking to facilitate innovation transfer. These initiatives include steps to streamline university procedures and strategies for setting up new companies. She emphasized, in this regard, the role that SBIR can play in facilitating technology transfer from the university to the market.

The SBIR program's Phase I grants, normally limited to $100,000, are competitively awarded to small businesses. They can also be a source of funding for university faculty to conduct feasibility research intended to establish a research idea's scientific and commercial promise. In addition, STTR provides awards to

---

[18]Companies of all sizes can participate in ATP-funded projects. To date, 68 percent of ATP awards are to small businesses or to joint ventures led by a small business. One NRC recommendation for ATP was to retain the valuable synergy offered by cooperation between innovative small firms and large companies. It noted that "large companies bring unique resources and capabilities to the development of new technologies and can be valuable partners for technologically innovative companies new to the market." See National Research Council, *The Advanced Technology Program: Assessing Outcomes*, 2001, op. cit., pp. 95–96.

researchers working with university researchers as sub-contractors.[19] SBIR and, to a lesser extent, STTR are proven mechanisms enabling collaboration between university faculty and small technology companies. University faculties have successfully used SBIR to found and fund their spin off companies to develop and commercialize technology resulting from their academic research. Phase I winners also compete for Phase II grants, which are intended to *develop the scientific and technical potential* of the research idea. Phase II grants are larger—normally up to $750,000. SBIR also has a final Phase III where grant recipients are expected to obtain additional funds. In the Defense Department, this can be procurement funds. In other agencies, it is often private investors, or the capital markets, that help *commercialize the technology*—there is normally no SBIR funding for this stage.[20]

SBIR has already proven to be an effective tool to solicit ideas and technologies from small firms to help fight the war on terror. Carole Heilman of the National Institute for Allergies and Infectious Diseases highlighted the importance of SBIR grants to the small business community, which "really rallied after 9/11." Within about a month, she said, the National Institute for Allergy and Infectious Diseases at NIH had put out a solicitation to the small business community, detailing what was needed to meet specific agency missions. This drew about three hundred responses within a month: "It was a phenomenal expression of interest and capability and good application, with extremely thoughtful approaches," she said. The ability of the SBIR solicitation to draw on the ingenuity of the vibrant U.S. small business community is one of the strengths of the SBIR program approach.

Industry Consortia

New technologies in the war on terror can also be facilitated through the best-practice lessons gained from experience in consortium-based cooperation among firms. Kenneth Flamm of the University of Texas described some best practice lessons in assessing the contributions of the Sematech consortium, which he described as the highest profile government-industry consortium in the United

---

[19]STTR is a highly competitive program that reserves a specific percentage of extramural federal R&D funding (2 percent) for award to small business and nonprofit research institution partners. Small businesses must be American-owned and independently operated, but the principal researcher need not be employed by the small business.

[20]Some agencies have adopted a Phase II enhancement. Called Phase II-B at NSF, it is designed to reinforce successful projects. For a detailed description of the SBIR program, see National Research Council, *The Small Business Innovation Research Program: Program Diversity and Assessment Challenges,* C. Wessner, ed., Washington, D.C.: National Academies Press, 2004. See also National Research Council, *The Small Business Innovation Research Program: Challenges and Opportunities,* C. Wessner, ed., Washington, D.C.: National Academy Press, 1999.

States.[21] These best practices include the invention of the industry roadmap and the emphasis on industry leadership to direct the partnership and reach consensus on research priorities.

Dr. Flamm traced the origins of Sematech to the early 1980s when the U.S. semiconductor manufacturers had lost their lead in manufacturing.[22] With Japan's substantial investments (beginning in the 1950s) in semiconductor materials, equipment, and manufacturing technology paying off in the late 1970s and 1980s, U.S. firms found themselves falling behind in manufacturing yields and market share. In this environment, the idea of a government-industry partnership arose as a mechanism to sort through the challenges and proposed a unified strategy, he said.[23]

Dr. Flamm noted that Sematech's designers decided to emphasize strategies that would improve the equipment and materials used by U.S. producers. Under the guidance of William Spencer, the consortium began to focus on reducing the time between new technology nodes and speeding up the flow of technology. It also developed a technology roadmap to provide coordination points for joint efforts in pre-competitive research. This roadmap, he said, has been judged a "hugely" important and innovative model for global technology formation, and has since been adopted internationally. Sematech itself is now an international consortium. One of its remarkable features, according to Flamm, is that companies that otherwise compete against one another in the marketplace voluntarily assemble every year to try jointly to identify technological obstacles and make plans to overcome them collectively. He suggests that "this is a new and totally unique phenomenon . . ." and perhaps "one of the lasting contributions of Sematech—a template for a new form of R&D collaboration."[24]

---

[21]In an industry R&D consortium, a certain portion and type of a participating company's R&D is funneled into a separate organization where it is carried out collectively and where the research results are shared among the member firms. In a consortium, firms can lower R&D costs or increase R&D efficiency while continuing to compete privately through their own product related R&D programs. The role for government in this partnership is to legally enable the inter-firm cooperation and, where appropriate, contribute funding and/or research facilities to advance research on technologies of common interest. See National Research Council, *Government-Industry Partnerships for the Development of New Technologies: Summary Report*, op. cit. p. 10.

[22]See Jeffrey T. Macher, David C. Mowery, and David A. Hodges, "Semiconductors," in National Research Council, *U.S. Industry in 2000*, David C. Mowery, ed., Washington, D.C.: National Academy Press, 1999.

[23]For a listing of events leading up to the decision to create Sematech, see Andrew A Procassini, *Competitors in Alliance: Industrial Associations, Global Rivalries, and Business-Government Relations*. Westport, CT: Quorum Books, 1995. For an overview of current programs to support the semiconductor industry in Japan, Taiwan, Europe, and the United States, see National Research Council, *Securing the Future: Regional and National Programs to Support the Semiconductor Industry*, C. Wessner, ed,, Washington, D.C.: The National Academies Press, 2003.

[24]See the presentation by Kenneth Flamm in Panel III of this volume.

Dr. Flamm also referred to the widespread agreement that Sematech played a significant role in the resurgence of the U.S. semiconductor industry.[25] While acknowledging that inadequate data collection makes it currently impossible to determine the extent to which Sematech was responsible for the resurgence, there is little doubt, he added, that the U.S. semiconductor industry did come back and that many U.S. firms are today on the leading edge of manufacturing. More telling, he said, was that Sematech is widely credited in Japan with considerable accomplishment, with the Japanese copying the structure of Sematech in their semiconductor strategy.[26] Further, he noted, the success of the consortium is also revealed by the willingness of Sematech's members to increase their funding for the consortium to about $140 million a year after the government subsidy disappeared.[27]

Moreover, the advance of productivity in the semiconductor industry, which accelerated dramatically after the major strategies of Sematech were installed, is linked to the simultaneous upsurge in U.S. productivity in the mid-1990s—a phenomenon documented by Dale Jorgenson.[28] "There was a direct link, noted Dr. Flamm, between this improvement in the pace of introduction of semiconductor technology and the improvement in the aggregate macro-economic performance of the U.S. economy."

While Sematech itself was designed to promote the competitiveness of a strategic U.S. industry through the joint development of new platform technologies, its best practice lessons are broadly relevant to the design of effective partnerships in the war on terror. Sematech demonstrated that industry roadmaps could help accelerate the rate of innovation by coordinating research among multiple actors and by setting the pace of market competition. And as Bill Spencer also noted at the conference, industry leadership, and cost sharing—important features of the Sematech model—provide the experience, expertise, and motivation required for partnerships to succeed.

---

[25]For a review of the evidence, see Kenneth Flamm and Qifei Wang, "Sematech Revisited: Assessing the Consortium's Impacts on Semiconductor R&D," in National Research Council, *Securing the Future: Regional and National Programs to Support the Semiconductor Industry,* op. cit., pp. 254–281.

[26]Ibid.

[27]Federal funding for Sematech ended in 1996.

[28]Dale Jorgenson has tracked the relationship between advancing semiconductor productivity and U.S. productivity. See D. Jorgenson and Kevin Stiroh, "Raising the Speed Limit: U.S. Economic Growth in the Information Age," in National Research Council, *Measuring and Sustaining the New Economy: Report of a Workshop,* D. Jorgenson and C. Wessner, eds., Washington, D.C.: National Academy Press, 2002.

PARTNERING FOR HOMELAND SECURITY— NEW CHALLENGES

Several conference participants highlighted key organizational, legal, and resource challenges facing the development of new public-private partnerships for homeland security.

S&T at the Department of Homeland Security

Congressman Boehlert noted that the new Department of Homeland Security (DHS) must be structured to draw together the science and technology expertise, funding, and policy attention needed to win the war on terror. Partnerships between government, industry, and academia will be necessary, he noted, because they can bring together the expertise needed to address the multiple dimensions of the homeland security threat.

In turn, William Bonvillian, of the Office of Senator Lieberman, outlined planned DHS steps to foster government-industry interaction. Among other initiatives, he noted, the new department is expected to:

- Create the position of Undersecretary for Science and Technology to ensure high-level policy attention.
- Establish federally funded research and development centers (FFRDC) to increase capacity in the area of risk assessment and risk management.
- Develop a clearinghouse to manage, identify, and evaluate technological opportunities that might be relevant to the agency's mission.
- Create an entity to encourage and sponsor technology transition.
- Create a DARPA-like entity to focus and accelerate research though government-industry partnerships and to leverage participation and cooperation across agencies.

Further to the last point, Larry Kerr of DHS added that the new agency plans to establish the Homeland Security Advanced Projects Agency (HSARPA) that would be "the systems equivalent of DARPA, but with many of the procurement issues and problems put aside." HSARPA would help "couple the research and development and testing and evaluation enterprises with actual entities—whether they are in the private sector or in academia—and the actual end users."

Addressing Liability, Regulation, and Intellectual Property Issues

Confusing liability, regulation, and intellectual property issues can be a serious impediment to effective public-private partnerships. Christina Gabriel of Carnegie Mellon University described how complex federal regulations (including new and changing rules on export controls), nonprofit tax law in some states, liability concerns, and the fear of legal disputes can limit university-industry partnerships. Effective technology transfer, she noted, requires that both partners hold similar objectives, communicate regularly, and build a bond of trust.

Referring to partnerships in bio-medicine, Kathy Behrens of RS Investment Management noted that questions regarding liability and regulation contribute significantly to the cost and time of development of many therapeutic agents. Stressing the importance of good communication among partners, she suggested that these issues be addressed thoroughly during the design phase of any proposed partnership.

Gail Cassell observed that the question of liability is extraordinarily complex in relation to human health and bio-threat agents, both because of the animal model rule[29] and the inability of firms to gather sufficient data to show safety and efficacy in humans. If anti-trust issues that preclude company consortia could be resolved, she noted, the resulting partnerships could provide working relationships that allow risks to be shared.

Responding to a question on whether product liability should be extended to cover software—an issue that potentially could impede partnerships for homeland security—Gordon Moore noted that the threat of excessive liability can slow innovation. He said that society had to decide on a balance that made sense. "Probably, liabilities will be pushed further than technical people would have wanted if they'd thought of it in the beginning," he noted.

Finally, with regard to intellectual property, Stephen Merrill of the National Academies noted that HHS Secretary Thompson had raised the possibility of abrogating the Bayer patent on the antibiotic Cipro if that measure were necessary to obtain an adequate supply of the antibiotic for an emergency. At the time, he said, some warned that this comment could have a chilling effect on companies' willingness to develop antibiotics, vaccines, and anti-virals. On this issue, Dr. Heilman said her work in the area of vaccines has shown that the government would have to set a policy environment that would not only nurture public-private partnerships but would provide strong intellectual property rights if firms are to be encouraged to undertake high-cost, high-risk, vaccine research.

Developing a Skilled Workforce

Workforce issues related to bio-defense remain a major concern, according to Carole Heilman. Not only are there insufficient numbers of people now trained in the microbiological and immunological sciences, but there are few incentives to attract them away from other important research. She estimated that there are

---

[29]The FDA's "animal model rule" is the principal approach to show scientific "proof of concept" for a candidate drug or vaccine that is under development as a countermeasure to a potential agent of bioterrorism.

INTRODUCTION

"maybe three people" in the United States with expertise in plague and anthrax. She also said that the nation lacked sufficient numbers of *in vivo* biologists—people trained in whole-body physiology. The need for experts in veterinary science is also a major issue, she said, not only to help establish infectivity models, but also address animal diseases and agro-terrorism.

To increase this small pool of expertise, she noted that the Department of Health and Human Services has established targets for training in bio-defense and intends to encourage partnering among the National Institutes of Health, regional public health service systems, and the Centers for Disease Control and Prevention.

The Environment for Innovation

Drawing together these concerns, Gordon Moore noted that "some deep partnerships between government and industry . . . are *implicit* rather than explicit." Policies to promote education and training, and regulations, anti-trust laws, and intellectual property laws that govern how organizations behave, he said, help create an environment for innovation and value creation. The structures of taxation, fiscal policy, and monetary policy also frame the context for partnerships. Together, he noted, these rules have made the United States the most productive place in the world to create technological innovations and transfer their value through the marketplace.

PUTTING IT TOGETHER

The importance of a systems approach to harmonizing disparate technologies and social cultures together in the war on terror was an important undercurrent to the conference discussions. Christina Gabriel noted that in addition to a sound policy agenda, good goals, and quality leadership, a partnership program must also possess the right operational features. These would take into account how people in the program interact and what incentives invite people naturally to work toward the goals of the program. Relatedly, Ron Sega of the Department of Defense described how developing complex weapons systems required developing effective networks that incorporate different systems so that they can interface effectively.

Channeling the vast amount of information that government agencies and other organizations must deal with in addressing the terrorist threat is an enormous challenge, noted Anne Altman of IBM. She identified three facets of this challenge: The first, she said, was the need to develop an *integrated information architecture*. Success here depends largely on organizing information lines and applying a common information strategy across the missions of various agencies. The second challenge is to create *partnerships to collect and manage information* both within government and among academia, businesses, and citizen groups.

Partnerships need to draw on the combined expertise of government, business, academic institutions, and citizens, she noted, since each member of this partnership brings unique information and abilities for optimizing homeland defense decision making. Finally Dr. Altman identified a need to *implement technologies and policies that ultimately enhance the government's ability to partner and achieve its missions.* "I think technology is key to cementing the partnership," she said. "We believe that it is the underpinning of open, standards-based architecture, allowing communication between various systems."

Integration will be a key challenge for the new Department of Homeland Security, whose six primary missions are to be accomplished by twenty-two constituent organizations. Jim Turner of the House Science Committee related the cautionary tale of the Department of Energy, which was created in 1977 in response to the Arab oil boycott. There was, he said, an "unnecessary amount of diversity in the agency," which "threw a lot of disparate problems together." As a result, "two things happened with DOE: From day one, the top management couldn't think about R&D." and second, "the DOE did not achieve the objective of weaning America from dependence on foreign oil."

Mr. Turner noted, however, that there are significant differences today, which are cause for optimism for the future of DHS missions. The nation has had over 25 years of experience with successful partnerships, he said, including Sematech, ATP, and SBIR, and the benefit of review and analysis of their best practices. Research on public-private partnerships led by the National Academies, he concluded, will help us understand how partnerships work, and these lessons can contribute to the nation's success in the war on terrorism.

This war on terrorism presents unique challenges. The strength of the United States in science and technology must be used to make the nation less vulnerable to future terrorist attacks and to reduce the risk and potential impact of such attacks. Speed is important. We need solutions to these vulnerabilities as soon as we can find them. This acceleration in the development of new technologies and new anti-terrorist products can best be done through partnerships between industry, government, and universities. The partnerships described here are much more effective than the "Silo" approaches to finding solutions. And they are also likely to prove to be much faster in mobilizing the strengths of the private sector to meet national needs.

II

PROCEEDINGS

Welcome

Bruce Alberts
National Academy of Sciences

Dr. Alberts opened the proceedings with a welcome and a brief introduction to the three Academies: the National Academy of Sciences, the National Academy of Engineering, and the Institute of Medicine. The three institutions have a total of some 5,000 members. The operating arm of the entity, formed during World War I, is the National Research Council. The combined Academies produce more than one report every working day, most of them for the federal government.

Noting the breadth of the response by the Academies to the September 11 attacks, he singled out the "rather heroic effort" on the part of some 160 volunteers who produced a major report released late in June 2002, titled *Making the Nation Safer: The Role of Science and Technology in Countering Terrorism.*[1] That study was initiated in the same room as the current workshop was being held, on September 26, when some 35 scientists and security experts gathered to recommend how the Academies could contribute to the "new, changed world." That gathering, he said, led to the current workshop, as well as "at least 50 others underway currently, attempting to bring the great strength of science and technology in this nation to bear on protecting the United States."

He noted that he had just returned from a week in Uganda, which prompted him to emphasize that terrorism is a worldwide problem. He reported that science

---

[1] National Research Council, *Making the Nation Safer: The Role of Science and Technology in Countering Terrorism*, Lewis M. Branscomb and Richard D. Klausner, eds., Washington, D.C.: The National Academies Press, 2002.

and technology in the United States was highly respected around the world, and that he had received "undue" praise for it while abroad. Scientists and engineers, he said, can have a profound and beneficial effect around the world by strengthening the scientific capacity of other nations.

He mentioned a major effort of the National Academies to "help develop in every nation the kind of capability that scientists have in this nation to advise their governments both on what we call policy for science—how to make science effective for meeting national needs—and science for policy—how to make wise decisions about the environment, water, health, and the future.

"All those decisions need to be made at the national level," he said. "They can't be made unless there are strong institutions connected to government, yet independent from government, like this one." He suggested that one of the major missions of the Academies was to "spread more rationality throughout the world through science and technology."

Introduction

William Spencer
International SEMATECH

Dr. Spencer said that his last job had been as chairman of SEMATECH, which seemed "almost a lifetime ago now." He noted his pleasure in serving over the last four years as the vice-chair of the STEP study of public-private partnerships, which has included partnerships spanning the sectors of industry, government, and academia. The purpose of initiating the current study, he said, was to examine functioning partnerships and extract lessons about how such partnerships might be strengthened.

He noted that at the outset of the study, he had tried to learn when the first such partnership had occurred in history, and had found, to his surprise, that "they have probably been around since before history." He said that a historian of technology had posited the likely scenario of an early human ancestor who could make stone tools more effectively than the others of his tribe, and therefore received a share of the hunt even though he did not participate.

In view of the long history of partnerships, therefore, the STEP Board had not tried to determine whether they should or should not exist; the Board simply accepted that they do exist in many forms and for many functions. Nor did the board try to compare partnerships according to degree of success, either in the United States or abroad. Dr. Spencer said that after 10 years at SEMATECH, he had learned that determining whether a consortium or partnership is "successful" is difficult to do.

Instead, the study group chose to try to understand what kinds of activities had been supported by partnerships, and which of those had achieved their objectives. They examined different kinds of partnerships, including consortia, competitive awards to innovative small firms, and relationships in science and technology clusters. He noted that he had just visited the new science park outside Sandia Laboratories in New Mexico, which features many public-private partnerships, and that it was showing substantial growth and accumulation of new companies even in the face of a subdued national economy.

He said that he would go over some of the main lessons he had absorbed during the time of the group's study, and cautioned that these lessons were "highly influenced by the time I spent at SEMATECH, as well as what the STEP board learned from our meetings." He noted that the study, which focused on current and proposed partnerships, had held about a dozen and a half major symposia and workshops, and had produced eleven reports over the past four years.

In one study he cited, STEP examined a difference in the way R&D in the biotech and pharmaceutical areas is funded, as opposed to the way R&D in the computer area is funded. He said that the lesson there, on which the participants at the meeting agreed by consensus, was that funding for the physical sciences was essential for continued advances in health sciences and life sciences. "I was pleased to see that the President's science advisory council has picked that up and is proposing a significant increase in physical sciences." He said that the advances in our life and health science programs would likely slow without new instrumentation, better measurement techniques, and sophisticated light sources, such as those produced at Lawrence Berkeley lab and Argonne National Laboratory, where it is possible to study the crystals of complex proteins.

A second study he described as significant was a comparative examination of partnerships in Japan, Taiwan, Europe, and the United States in the semiconductor and electronics industries. The group had begun its study with the help of "a small paper by Kenneth Flamm, about partnerships in Japan 5 or 6 years previously," and extended that study to the semiconductor industry of Taiwan and China and the many partnerships in electronics in Europe.

Lessons from the STEP studies

From those studies, he said, the group had learned several lessons. The first was that in a partnership between government and industry, it is essential to have a clear and measurable set of objectives. "If you don't know where you're going," he said, "almost any direction will get you there." These objectives need to be established, measured, and then reported on regularly. He also said that the objectives should be focused as closely as possible on generic or precompetitive work rather than on products for the commercial market. "We absolutely stayed out of products and any processes for products at SEMATECH," he said, "and I believe that consortia or partnerships with government to develop products are not going to work."

A second lesson, he said, was that even though those objectives need to be set and measured, the group found it was necessary to maintain flexibility. When

SEMATECH began, in 1987 and 1988, there was a strong belief that the founders of the consortium were primarily trying to renew semiconductor memory capability in the United States. That was not the explicit goal, said Dr. Spencer, but even so, the consortium was able to make major changes within 2 years of it's start, which resulted in contributions to the health of the industry as a whole. He added that the consortium quickly found that it needed a clear roadmap of where it was going, a practice that has been adopted over the last 10 years, not only by the semiconductor industry but now by many other organizations.

A third lesson was that a partnership needs to be led by the very best people in the industry. In the case of SEMATECH, he said, that was true "from the top of the company down to the people who worked in the consortium. Quality leadership and quality people participating is a rule that needs to be followed."

He then borrowed a mathematical expression to say that these conditions were necessary but not sufficient to bring positive change. "I don't know of a close set of sufficient conditions we could write down," he said, "that would ensure that a partnership will succeed." These lessons need to be followed, he said, but doing so does not guarantee success.

He closed by praising the commitment and capacity of the steering committee, which he described as "extraordinary even by NRC standards," and the staff, noting that the present workshop would be the last of this particular series of meetings on government-industry partnerships.

Panel I ———————————————————

Partnering to Meet the New Security Challenge

INTRODUCTION

Sean O'Keefe
National Aeronautics and Space Administration

Mr. O'Keefe "set the stage a bit" by saying that both the primary focus of this workshop—partnering against terrorism—as well as the topic of this particular panel discussion—partnering to meet new security challenges—would cover a range of different questions, and that they represented a "very contemporary topic of public debate in Washington, as well as throughout the academic community and industry." He said that trying to narrow the scope of those discussions would be part of the challenge of workshop participants.

He opened by sketching the context "of how we at NASA have viewed this [challenge] since September 11," which meant "a rather dramatic alteration" toward new efforts to "focus dominantly on homeland security and move away from some of the historic charter mission objectives." The agency had begun to think of "how you employ those extant, current assets and capabilities in different ways to meet what are now a very focused set of challenges."

Those challenges, he emphasized, did not arise suddenly on September 11, but prior to that time, when a series of earlier terrorist events demonstrated the new reality the nation would be forced to confront. Certainly, he said, members of the Hart-Rudman commission and others before them had identified the seriousness and extent of international terrorism.[2]

---

[2]The U.S. Commission on National Security/21st Century released its *Roadmap for National Security: Imperative for Change* in January 2001, some 8 months before September 11. The commission, chaired by former U.S. Senators Gary Hart and Warren Rudman, was an independent panel created by Congress to conduct "the most comprehensive review of American Security since the National Security Act of 1947 was signed into law over 50 years ago." The report urged creation of a new "Homeland Security Agency, and warned, "States, terrorists, and other disaffected groups will acquire weapons of mass destruction, and some will use them. Americans will likely die on American soil, possibly in large numbers."

NASA's Approach to Terrorism

The approach taken at NASA was to look clearly at its missions in light of this new reality. In the case of its aeronautics mission, for example, the agency asked whether its activities could be adapted to help deal both with terrorist tendencies and the challenges of aviation on any given day. These challenges included averting or avoiding collisions with inanimate objects of any sort, and using capacities, some developed by public-private partnerships, to reduce the danger of weather-related incidents.

This broadened agency focus meant paying heightened attention to aviation safety in general. Within the last month, he said, the agency had stepped up its attention to advancing the kinds of technology related to safety. Operationally this included (1) demonstrating a technology developed in partnership with several applications from industry to devise a specific methodology, and (2) testing that technology aboard an environment equivalent to that of a commercial airliner. This technology was designed to provide a pilot with an early alert of a potential collision, to repeat that alert several times, and to take evasive action automatically in case that action was not taken by the pilot. With this technology in place, the remaining challenges are operational—the task of taking such a capability and making it available to the real world of commercial conditions in a way that is not intrusive. Mr. O'Keefe emphasized that parts of this task involve debates and conflicts that had been easier to avoid previously. Now, he said, "there is an imperative to address them."

He then introduced the first panelist, Congressman Boehlert, as a person who had been "not only incredibly supportive, but has led the way on science and technology objectives in his 20 years in Congress."

PARTNERING FOR CYBER SECURITY AND INFRASTRUCTURE PROTECTION

Congressman Sherwood L. Boehlert (R-NY)

Congressman Boehlert began by listing what he called "key points of agreement on what we are here to discuss today":

1. Homeland security has to be a primary focus of activities across the federal government.

2. A Department of Homeland Security (DHS) is needed so that one agency is especially focused on the country's security needs and so that security-related activities can be better coordinated.

3. Science and technology must be essential elements of the work of a Department of Homeland Security, and any homeland security strategy.

4. Cyber security is one of the critical areas to address in homeland security science and technology efforts.

5. The structure of the Department of Homeland Security has to reflect the significance of science and technology and enable the Department to attract the expertise needed to oversee science and technology.

6. Science and technology activities for homeland security need to be carried out in partnership between government, industry, and academia.

7. The dispersed nature of the homeland security threat requires that government and industry work together more closely than ever before.

These beliefs, he asserted, were shared by both parties, in both houses of Congress, at both ends of Pennsylvania Avenue, and across the ideological spectrum.

He said further that while these seven conclusions may now seem self-evident, they had not been self-evident in Washington just a few months earlier. As an example of the significant movement since then, he said that when the House Science Committee began its work on homeland security legislation at the end of June 2002, it had to overcome resistance from the Administration to create an undersecretary for science and technology in the new department. By the time the bill came to the house floor at the end of July, however, the Committee's proposal had been endorsed by Gov. Tom Ridge,[3] and had since been backed by the President's Council of Advisors on Science and Technology (PCAST). The Committee's science and technology focus was also duplicated in the Lieberman bill in the Senate. Moreover, he said, and more tellingly, the Senate Republican counterproposal, crafted with the White House, also maintained the Committee's science and technology structure.[4]

The Undersecretary for S&T and the Need for Partnerships

That is a big change since June, he said, and a significant change. The debate over whether to have an undersecretary for science and technology was not a struggle over bureaucratic minutiae; the issue was whether the department was going to have a clear science and technology focus, with responsibility and accountability concentrated in one person with the expertise to assemble a credible staff and to oversee research and development. The existence of the undersecretary and a secretariat, he said, would give R&D the "heft" it was sorely lacking in the original bill. This condition also meant that the department's R&D functions and budget would be more than the sum of small, disparate pieces transferred into the new department from other federal agencies. "And that has to be the case," he said, "if we are to succeed in the war on terrorism. As I often say, the war on terrorism, like the Cold War, is going to by won in the laboratory as much as on the battlefield. So that laboratory has to be adequately stocked."

---

[3]Governor Ridge was later named Secretary of the Department of Homeland Security (DHS).

[4]The first Undersecretary of Homeland Security for Science and Technology, Charles McQueary, was sworn in on April 9, 2003, at the National Academies building, 2101 Constitution Ave., Washington, D.C.

How are we going to ensure this, he asked. Obviously, he answered, not by relying on an in-house federal capability: "That wouldn't be just infeasible, it would be unwise." As in every other area of R&D, the federal government would have to work cooperatively, in partnership with academia, industry, and the states. That is especially true in the area of homeland security, he said, because the problems are so varied; the needed expertise must be gathered from almost every discipline and the results of any R&D will have to be applied as much by the private sector as by the government. Expanding on that point, he reminded his audience that the September 11 attacks had targeted not only public buildings—nor had the anthrax attacks targeted only public buildings: "Every individual and every sector of the economy is at risk."

A prerequisite to the involvement of industry, he said, is that products developed to thwart terrorists will have to meet the needs of private entities and succeed in the private marketplace. And yet such products must be developed even without assured demand. "If there was ever an endeavor that cried out for public-private partnerships," he said, "it is the research and development related to homeland security. Here is a case in which the government cannot carry out its most basic mission of providing security without the cooperation of the private sector. And here is a case in which the private sector will quickly need a range of products on which the market has never before put a premium. This is a classic case of market failure that calls out for government involvement."

A Change in Thinking About Security

Congressman Boehlert said that it was striking to realize how quickly the thinking about security had changed. For example, he said, his staff had been at a meeting the previous week in his home district with the real estate round table. The round table represented major developers of office buildings, malls, and other commercial properties. Yet the subject that received the most attention during the meeting, he said, was homeland security. One of the participants had come from the White House, and interest in homeland security was so high that the session ran well beyond its scheduled ending. "Can you image that subject even being on the agenda before September 11?" he asked. "So how do we craft an R&D program that meets the needs of commercial real estate developers? That's a new kind of question."

The Congress was well aware of this change, he said, and the need for cooperation, even if that awareness had not yet been fully articulated. He cited the cyber security legislation his committee hoped to send to the President within days as a good example of this new state of affairs. Until the attacks of September 11, he said, it was difficult to get many members of Congress to focus on the cyber threat. And the members of Congress had hardly been alone in their lack of concern. The marketplace for software, for example, favored not security but speed, ease of use, and low prices. Software developers who focused on security did so at their own economic peril. That had now begun to change, he said, just as cyber security had become a hot topic on Capitol Hill. The software market was beginning to send signals that security had become a desirable feature of software.

He noted an additional problem: Cyber security was not an area where the best course of action was known, lacking only the wherewithal to proceed. The bigger problem was that technical people did not yet know enough about designing more secure computers and networks. He referred to Bill Wulf, president of the National Academy of Engineering, who had described the cyber security paradigm as a Maginot Line defense, after the notoriously porous perimeter that failed the French during World War II. "Clearly," said the Congressman, "we need some new ideas. But how do we get them? The answer, as I've already implied, is through creative partnerships."

The government has to be involved, he said, because improving cyber security required additional basic research, and it also required greater support for students in order to attract new people to the computer security field. Academia has to be involved because much of the expertise in this area resides in colleges and universities, which also have the capacity to educate a cadre of computer security experts. Finally, industry has to be involved, because private firms have perspective on what is needed in this rapidly evolving field, and because advances in computer security must be able to succeed in the private marketplace if they are to have a broad impact.

He said that the House Science Committee had assembled a bill called the Computer Security Research and Development Act, H.R. 3394. The bill would create a variety of new programs at the National Science Foundation (NSF) and the National Institute of Standards and Technology (NIST) to attract more researchers to the field of computer security and encourage them to come up with more innovative ideas. It also proposed several new programs to fund partnerships between universities and industry.

The NSF-funded partnerships were to be approved through traditional peer-review processes and would focus on basic research. The NIST-funded partnerships were to be selected by program managers, using a process modeled on that of DARPA that focused more directly on problems identified by industry.

The Congressman said he introduced the bill in the House in December 2002, while Sen. Ron Wyden (D-Ore) introduced it in the Senate. He noted that the bill passed the House by a vote of 400–12 in February, which he said was "an impressive vote" for an $880 million new program, and he noted that the Senate was likely to pass a slightly revised version. He called this "a rapid response for Congress, a warm and overwhelming endorsement for the concept of partnering against terrorism."[5]

---

[5]The Cyber Security Research and Development Act became law on November 27, 2002.

How the Homeland Security Bill Supports Partnerships

He noted that the bill supported the concept of partnerships in several ways. Most significantly, the House bill would create a clearinghouse to ensure that individuals and companies with ideas related to homeland security did not get lost in the maze of federal agency jurisdictions. The idea for the clearinghouse, based on experience with the Interagency Technical Support Working Group, was to create a single point of entry into the federal government for people in the private sector with ideas or products that might help enhance homeland security. The need for such an operation, he said, became "painfully obvious" in the wake of the anthrax attacks when the government was deluged with suggestions. Several House committees, including the Science Committee, pressed for creation of such a point of entry, which indicated the degree of concern. "I don't think there's much doubt in Congress that partnerships are a key element of any R&D strategy for homeland security," he said. "And I'm sure the excellent work the STEP board has done in recent years in describing how partnerships can work, along with your discussions today, will help shape that strategy."

A difficult question that remained, he said, was not whether to promote partnerships, or whether to have a Department of Homeland Security, or even how to fund such a department. It was how to maintain the traditional ability of scientists and engineers to publish and communicate about their research without jeopardizing homeland security. He cited discussions among government agencies about how to categorize and regulate information that is "sensitive but unclassified," and whether to develop new restrictions on the conduct of research by foreign-born faculty and students. "How to strike the proper balance between the openness research needs and the security the nation needs," he said, "is not obvious."

The magnitude of the problem is illustrated, he said, by analogous efforts by private companies to limit the flow of scientific information produced by their partnerships with universities—a similarly difficult balance between openness and security. As a member of both the House Science and Intelligence Committees, which "tend to err in opposite directions on this issue," he said that he knew how tough an issue this was going to be. "All I can say is that we have serious thinking to do, and the balance is going to have to be constantly recalibrated. We do want to open up public dialog on the issue. I know the National Academies will want to do the same, which is why I dangle this perplexing and unresolved matter before you."

The fact that the issue of openness had come to the fore, he said, was one more indication of how much the world, or at least our understanding of the world, had changed since September 11. He closed with the puzzlement expressed long ago by Shakespeare: "Oh brave new world that has such people in it."

"We're all going to have to work together," he concluded, "if those people are going to be held at bay."

CAPITALIZING ON THE NATION'S RESEARCH PORTFOLIO

Gordon Moore
Intel Corporation

Dr. Moore began by acknowledging that he was not an expert on terrorist threats, but that he had learned, through his work chairing the GIP Committee, some lessons about partnerships. He asserted that technology was going to be a critical component of the country's response to the post-September 11 security challenges, and that partnerships were going to be an important way to focus that technology on those problems. The United States, he said, had "the best and broadest science and technology in the world." And our job is to see how we can "make that intersect the problem space we've discovered in the last year."

He suggested that the country was not attaining its potential in the application of science and technology to counterterrorism. The challenges of collecting, analyzing, and insuring intelligence, for example, had not been addressed in ways that could be utilized by the government. Likewise, federal agencies were making only slow progress in deploying effective explosives detectors in the nation's airports.

He said that the STEP board had studied for more than over four years how partnerships could be used and, he thought, had probably helped policy makers better understand the role that partnerships could play. Although the committee had not until the present workshop specifically addressed the role of partnerships with respect to terrorism, it had addressed a number of technical areas that would be relevant to questions regarding weapons of mass destruction and other challenges.

He cited several areas where sufficient technical knowledge may rely on the use of government-industry partnerships, including new instrumentation to detect radiation at large distances, perhaps making use of technology first developed for gamma ray astronomy. A second important area requiring more technology was rapid identification of bio-agents, including vaccines, antibiotics, and anti-virals, some of which already existed. He noted that he served on the board of a small company that had found a drug that was originally produced for a certain disease but now also showed significant ability against various poxes, giving hope that it would find practical use against smallpox. "Some of these things exist," he said. "Clearly, partnerships can help put them in a position to be much more useful."

The Continuing Need for Long-term Research

He said that the proposed structure in the new Department of Homeland Security (DHS) for science appeared to have high potential value. He approved in particular of PCAST's recommendation that this structure contain a DARPA-like quality that supported "high-risk, far-reaching research." In his view, he said, DARPA over the years had been a successful mechanism for quickly addressing complex problems, "and the extension of that idea could be extremely important."

He added that the ability to address complex problems depended on having a broad base of leading-edge science and technology. While the federal science budget had gone up over the past several years, some parts, particularly those related to basic physical sciences and engineering had actually dropped, especially during the 1990s. "To me," said Dr. Moore, "this is a problem. Science moves on a broad front. You can't move one area much faster than the rest, because there is so much interdependence."

He acknowledged that there are efforts to do some "rebalancing" of the budget, based on what is needed to carry all of science and engineering forward. He cited the example of advances in biology that depend on imaging—which, in turn, depends on some of the physical science. Similarly, nanotechnology, which is likely to figure in important ways in strengthening national security, will depend on a basic understanding of materials needed to make structures and measurements at dimensions never before achieved. Finally, he said, new devices that we anticipate from engineering research are likely to be important not only to particular problems related to homeland security but also to the economy.

He then turned to information technology, which he singled out for special attention. The government faces unique IT challenges because of its sheer size and complexity. Throughout the government, information technology had developed in "small pockets" that were not interrelated, as it did originally in most companies. Enabling these pockets to talk to one another is a considerable challenge, he said, but one that has to be faced. Dr. Moore said that he had worked with companies and other institutions that were trying to upgrade their computer facilities so as to take advantage of modern technology. Even though the challenges facing companies were far smaller than those facing the U.S. government, and even though companies were more flexible than government, virtually all of them concluded such projects with fewer features than they envisioned, having spent more in time and money than they had planned. In other words, he said, these are "tough problems." Even though industry is collectively learning to handle them, they will be a new challenge for government at the scale of what is required.

"The solution," he said, "requires partnerships. It requires the best minds, including the flexibility to include future technology; it needs an adequate budget, built with off-the-shelf software and hardware products; and it probably requires a short-term and long-term strategy that will be most effective in handling the security problems we've looking forward to."

He said that during its four-plus-year study of government-industry partnerships, the STEP board had discovered that government-industry partnerships can indeed work. "They've helped us to meet some of our national missions," he said; "they've been used effectively since the founding of our country; and our studies have identified some useful features in producing results. The appropriate thing now is to apply what we've learned to meet the challenges posed by terrorism."

DISCUSSION

Mr. O'Keefe commented on the observation that the September 11 attacks were not really made against buildings, suggesting instead that they were aimed at the morale of the American public. "When we talk about public-private partnerships and prioritizing research," he said, "one important consideration must concern the morale of the American public, as they behave economically and sociologically. Often as scientists, we focus on electrons and molecular biology, and we don't think about perceptions or psychology. But when we talk about war involving terrorism, the bottom line is the morale of the American people."

Congressman Boehlert thanked him for that observation, and asserted that the terrorists had made "no dent in the morale." Before 9/11, he said, whenever the word "terrorism" came up, the context assumed for that word had been the Middle East. "If I'd mentioned the need for a Cyber Security Research and Development Act before 9/11, I have had a hard time getting many people enthused about it." On September 10, he said, he doubted if he could have convinced even the principals in such a project to show up for a meeting. Congressional representatives wanted to know, "How does that impact me, how does that affect my constituents." And yet recently that same bill had passed by 400 to 12. Homeland security had united many people, he said, persuading them to "point in the same direction."

A questioner asked Mr. O'Keefe about the strategy of the new Department of Homeland Security—whether it would build up its own internal laboratories and grant programs, or focus on strengthening the relationships of existing agencies with industry and universities. He replied that while he could not speak with authority on the department's strategy, he could suggest from NASA's point of view that it had been "refreshing" to see the effort by department planners to capitalize on historical legacies of the last 50 years.

Capitalizing on Existing Capabilities

To be sure, he said, for students of organization theory, those years came with a checkered history. He said that it was his impression that DHS planners were capitalizing heavily on successes and trying to avoid the "potholes," while recognizing the deep cultural differences between agencies and departments. He said that the centerpiece of organizing philosophy was to capitalize on existing capabilities, rather than attempting to expand beyond them. This would include a general trend of capitalizing on public-private partnerships—a trend that might be challenged by elements of homeland security that lack experience with such partnerships.

Another questioner asked what could be done to bring science and engineering into the decision-making process, especially where non-scientists "might go overboard in paranoia." He cited the example of a section of Pennsylvania Avenue in Washington that had been closed for seven years, noting that the White House had already been fortified against a car or truck bomb on the avenue with 660 tons of steel, concrete, and laminated glass windows. Yet, he said, there were current plans for additional expenditure of $6.1 million to break up Pennsylvania Avenue and install gravel, and on the E Street side, which is farther from the White House, to spend $100 million to build a tunnel. The questioner asked how scientists and engineers could help prevent such unwise expenditures, including those that might restrict basic American freedoms.

Mr. O'Keefe agreed with the questioner that "we've figured out the most difficult way possible to prepare for possible circumstances," including some "amazingly silly approaches." He said that most were proposed in good faith, and that with the passage of time, "sobriety will set in."

A questioner suggested that consolidating the activities of other agencies for the new department might disrupt working relationships in all sectors. Referring to Mr. O'Keefe's earlier experience at OMB, he asked whether it would not be desirable to name a new associate director of OMB for homeland security and adapt the multi-agency approach of OMB to the new Department of Homeland Security.

Mr. O'Keefe agreed that such a move would have the advantage of cross-cutting perspective, but he said that a larger question concerned how Congress would choose to consider and dispose of requests for resources, so as to keep the process moving. That, he said, would be "more important than some organizational twist."

Ronald Stoltz of Sandia Laboratories said that his facility was actively involved in a bridging role with Lawrence Livermore labs in preparation for the new DHS. He said they were using existing capabilities, not building new ones. Dr. Stoltz then directed a question to Gordon Moore about partnerships in Silicon Valley, specifically the issue of whether product liability should be extended to cover software. He said that is was a large issue that potentially could impede partnerships for homeland security, and asked if STEP had considered it. Dr. Moore answered that the committee had not yet considered it, but agreed on its importance. "If liabilities get extended too far," he said, "it slows innovation." He said that society had to decide on a balance that made sense. "Probably, liabilities will be pushed farther than technical people would have wanted if they'd thought of it in the beginning."

Mr. O'Keefe offered a similar view, saying that NASA had to deal with questions of product liability every day. "In trying to conquer challenges we've never had before," he said, "we often have no benchmark to calibrate the likelihood of success. We try technical forecasting, but it really comes down to risk management. That can be the fastest way to stifle innovation."

Panel II ———————————————————

Best Practice Examples of Public-Private Partnerships

INTRODUCTION

Arden Bement
National Institute of Standards and Technology

Dr. Bement noted that the first panel had established the basis and rationale for public-private partnerships associated with the nation's new security challenge. The current panel, he said, would take the issue a step further, to best practices and examples of public-private partnerships.

He noted that NIST was effectively engaged in matters of homeland security, including cyber security, counterterrorism, and critical infrastructure protection. Homeland security had now been designated as one of four strategic thrusts for NIST over the next decade, when the institute expected to increase organizational emphasis and investment.

He said that the two previous speakers had served as models of the kind of cooperation, dedication, and ingenuity necessary to prevail against the threat of terrorism. He praised Gordon Moore as a member of "the nation's pantheon of technologists," having "charted the semiconductor industry's course" in the information revolution and co-founded and led the company "that helped put the revolution into high gear." All along, he said, Dr. Moore championed the strategic importance of maintaining a strong national platform for innovation, which was now an asset fundamental to the successful response to the challenge of terrorism.

He noted that Congressman Boehlert came from the state that had borne the brunt of the "misery and devastation wrought by the attacks on September 11," and had become a forceful and effective advocate for policies that view the support of science and engineering research and their application as "investments in a better future." He said that Mr. Boehlert had been at the forefront of efforts to leverage the nation's science and technology resources in the fight against terrorism. Leveraging through partnerships and coordination, he said, will be key to how effectively we marshal our technological capabilities to counter the asymmetric threats of terrorism, a threat he called "unprecedented in terms of its dimensions and complexities."

As an example, he described the physical infrastructure at potential risk—the nation's collection of utilities, bridges, ports, water systems, airports, hospitals, plants, and factories, some 85 percent of which is privately owned. He pointed as well to our immense information infrastructure and its multitude of vulnerabilities, and to levels of emergency preparedness in 56 states, territories, and possessions, more than 3,000 counties, and tens of thousands of communities where 285 million citizens live. In all, he said, this presented a "systems problem of the highest order." The number of technical issues and scientific questions aside, he said, we face a gigantic organizational and operational challenge that can best be faced collaboratively.

NIST and Homeland Security

He said that NIST was supporting some 120 projects that address issues of homeland security, many of them characterized by collaboration. Some 75 of those projects, which had begun before 9/11, had been redirected. In the area of radiation standards, for example, NIST had already been developing standards, and had redirected its work to include development of standards for beta radiation. For DNA, the institute had been developing standards for analysis, and it shifted that work to the study of damaged DNA in order to assist in the identification of the victims of the World Trade Center collapse. In addition, NIST's Advanced Technology Project (ATP) supported companies that were bringing to life "embryonic technologies" through cost-sharing awards. Since 1992, ATP had obligated an estimated $270 million to companies and joint ventures pursuing promising commercial technologies that could be enlisted in the fight against terrorism. That partnership, he said, underscored the dual nature of many of the technologies that were now needed.

Many partners outside NIST were contributing to the homeland security work underway in NIST laboratories, with an emphasis on responding to measurement and standards-related needs. He cited three such needs:

(1) NIST was starting its investigation of the structural failure and progressive collapse of the World Trade Center, bringing together technical experts from industry, academia, and other laboratories and interacting regularly with the pro-

fessional community, local authorities, and the general public. NIST had also assigned a special liaison to families of first responders and families who had had members in the buildings at the time of the collapse. The investigation was part of a broader NIST response to the World Trade Center disaster.

(2) In concert with the World Trade Center investigation, NIST was conducting a multi-year research and development program that also engaged experts from the private sector, academia, and professional societies. The objective was to apply lessons learned and to use the results of this collaboration to provide a technical basis for improved building and fire codes, standards, and practices.

(3) NIST was also facilitating and supporting an industry-led program to disseminate information and technical assistance. This program was designed to provide practical guidance and tools to better prepare facility owners, contractors, designers, and emergency personnel to respond to future disasters, whether natural or human-caused. One challenge was to convey reliable information to people in the front lines of homeland security. The previous month, for example, NIST had issued the first comprehensive set of basic procedures for decontaminating protective clothing and equipment. This was being provided to personnel who were charged with responding to chemical, biological, radiological, or nuclear attack. The report consolidated recommendations and key information from many authoritative sources, including makers of synthetic fibers and protective equipment, fire departments, and government laboratories. This potentially life-saving reference was the result of collaboration between NIST, local fire marshals, U.S. fire administration, and the chairperson of the national Volunteer Fire Council. The manual was being made available without charge, and could be seen on the NIST web site.

The programs described under (2) and (3) both sought to gain lessons from the events of September 11 and to apply those lessons to new codes and standards.

Dr. Bement said that international collaboration was also critical in strengthening the nation's and the world's defenses against terrorism. The advanced encryption standard (AES), for example, was a result of such international cooperation, which also featured rigorous competition. The standard was selected from 15 algorithms submitted by cryptographers from around the world. The winning AES was developed by two cryptographers in Belgium and approved last December as the federal standard for civilian agencies; it was already finding widespread use in the private sector as well. The AES was designed to protect sensitive computerized information and financial transactions. He estimated that millions of people in both the public and private sectors would likely use this standard over its lifetime, which could span one to two decades, or even more. He said that Intel's chief security architect had described the selection of the encryption standard as a process that "should be held up as the model of industry-academia and government cooperation."

SEMATECH: ASSESSING THE CONTRIBUTION

Kenneth Flamm
University of Texas at Austin

Dr. Flamm said that he wanted to offer some comments about the contribution and lessons of SEMATECH that were based primarily on his own experience and opinions, rather than on rigorous analysis.

He began by sketching a picture of the semiconductor industry, to explain why economists—and the STEP board—placed so much emphasis on it. The semiconductor industry was now the largest U.S. manufacturing industry, measured by value added—the contribution to national output. He said that it may be a surprise "or even shocking" for some people to learn that it was even larger than the computer industry in the United States. Likewise, because value-added figures are those that relate to value originating within the industry itself, the semiconductor industry was larger than the automobile industry, which uses many components (such as semiconductors) that originate in other industries. As a single manufacturing industry, he said, the semiconductor industry was then approaching 1 percent of GDP; the entire manufacturing sector in the U.S. accounted for 17–18 percent of GDP.

Perhaps more importantly, semiconductors constituted a key input to other important industries—especially across the spectrum of information and communications technology (ICT). Semiconductors were probably the largest and most important input to all of the industries that made up the new realm of ICT. He said that this conclusion grew largely out of the work of Prof. Dale Jorgensen of Harvard, who had performed extensive research on the impact of semiconductors on GDP and ICT.[6] He also said that economists had substantially improved the quality of their statistics on the computer and communications industries. "Our understanding of the growth of the U.S. economy over the last two decades has been completely altered," he said, "by this revisiting of the basic numbers on the sources of productivity growth."

Price Performance of Semiconductors

He said that adequate data had made it possible to estimate what portion of the decline of computer and communications equipment prices—a measure of technological progress—was due to the price performance of semiconductors (see Figures 1 and 2). The calculations were laborious, he said, but in summary, he had found that about 40–60 percent of the decline in computer prices in 1998 was due to the decline in the cost of semiconductor functionality. The remaining 40 to

---

[6]Dale W. Jorgensen, "Information Technology and the U.S. Economy," Presidential Address to the American Economic Association, New Orleans, LA, January 6, 2001.

	Share of Total Price Change Due to Semis, 1998 (Percent)	
	Low	High
Consumer audio	23.1	31.5
Computers	39.0	58.5
LAN Equip	17.7	30.6
LAN Equip+ Switches (est.)	15.7	27.0

FIGURE 1 Role of semiconductors in computers and communications innovation. SOURCE: Aizcorbe, Flamm, and Khurshid (2002).

		Percent/Year
Microprocessors, Hedonic Index	1975-85	-37.5
	1985-94	-26.7
Intel Microprocessors, Fisher Matched Model, Quarterly Data		
	93:1-95:4	-47.0
	95:4-99:4	-61.6
DRAM Memory, Fisher Matched Model	1975-85	-40.4
	1985-94	-19.9
DRAMs, Fisher Matched Model, Quarterly Data		
	91:2-95:4	-11.9
	95:4-98:4	-64.0

FIGURE 2 Economic impacts: Decline rates in price-performance. SOURCES: Flamm (1997); Aizcorbe, Corrado, and Doms (2000).

60 percent of the cost decline was caused by innovation. He had made similar calculations for communications equipment, and found that 15 to 30 percent of price declines were due to price declines in semiconductors; for consumer audio, the figure was 20 to 30 percent. He added that the actual measure he had used was a quality-adjusted measure for the decline in price for a particular kind of equipment that made use of semiconductors.[7]

Despite the importance of the semiconductor sector, he said, "the data are awful." Given that this is the largest single manufacturing sector in the U.S., "you'd think that effort would be expended on collecting adequate numbers. We have better numbers on pork bellies and cheese and industrial fasteners than on semiconductors." He said there were many complicated and legitimate reasons for this, and an important one is the cost of collecting the needed data. Without adequate funding, the government relies on price data sold by market research companies. These data are undoubtedly cheaper, he said, but are not collected by the standards required by economists.

He then reviewed the origins and early years of the semiconductor industry, which "was basically a U.S. industry." The transistor was invented at Bell Labs in New Jersey, and the integrated circuit was developed at two U.S. laboratories independently. During the early years of the industry, in the 1960s, each of the major competing firms would design and manufacture not only semiconductors, but all the other ingredients needs for integrated circuits. They would develop their own materials, grow their own crystals, slice wafers, and design circuits. They would be engaged in front-end fabrication, the deposition and patterning of silicon wafers, and even the back-end activities of assembly and testing. But as the industry got bigger, it began to specialize. The semiconductor industry was one of the first to send its labor-intensive activities—the assembly and testing of semiconductors—offshore to Hong Kong, Southeast Asia, Japan, and Mexico.

Then in the 1970s, the integrated companies began to spin off their materials and equipment activities to firms that specialized in those activities. The 1980s saw the next step of specialization, most notably the "fabless" firms that did only chip design, leaving the fabrication operations to manufacturing firms downstream.[8]

With specialization came new kinds of coordination challenges. When all functions were done inside individual large firms, coordination was manageable.

---

[7]Ana Aizcorbe, Kenneth Flamm, and Anjum Khurshid, "The Role of Semiconductor Inputs in IT Hardware Price Decline: Computers vs. Communications," Federal Reserve Board, Finance and Economics Discussion Series, August 2002, http://www.federalreserve.gov.

[8]Taiwan Semiconductor Manufacturing Co. and a few other firms changed the semiconductor market by specializing in the manufacture of custom wafers under contract to chip designers. This freed the designers to concentrate on making and marketing the integrated circuits formed on the wafers to form microchips and helped spark an explosion of "fabless" microchip companies, such as those that populate Silicon Valley and the high-tech zone around Taipei.

Now, most firms found themselves involved in complex technology flows from different groups of vendors in different niches and countries. Only the very largest leader firms had the resources to coordinate the next generation of technology internally. This is a high-cost activity; those who attempt it have to accept that a certain amount of "leakage" will spill over to other firms.

A Review of Moore's Law

Another important event in the history of the semiconductor was "Moore's law." He recalled that Gordon Moore in 1965 published a paper in an IEEE journal suggesting that the density of devices on a chip would approximately double every 12 months for the next several years (see Figure 3). This meant essentially a doubling of the capacity of each integrated circuit. Ten years later, that prediction was still more or less accurate, and it became a de facto benchmark for predicting when the next generation of technology would come around. What started out as an optimistic prediction, said Dr. Flamm, began to take on a life of its own, becoming, by accident, a coordination device for the industry.

Around 1975, the rate of doubling slowed, and Dr. Moore raised his estimated cycle time to 24 months. This figure turned out to be pessimistic, as the doubling time rose to only about 18 months—a "middle ground between Moore One and Moore Two."

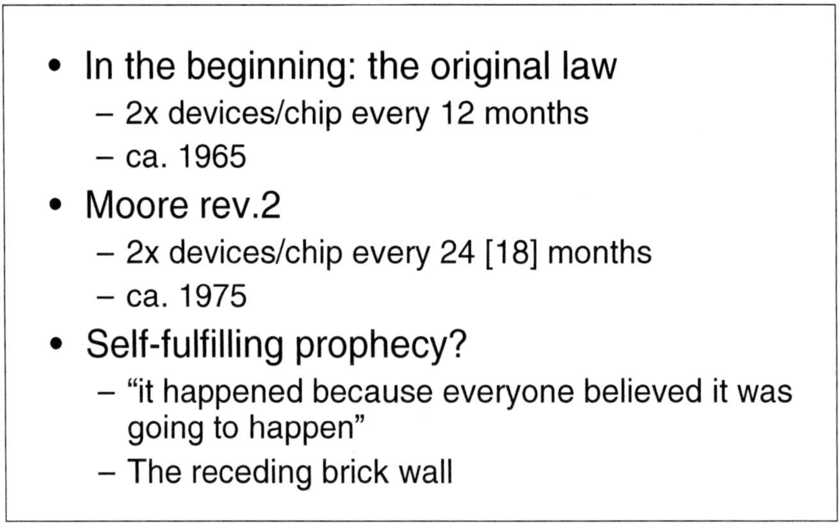

- In the beginning: the original law
  - 2x devices/chip every 12 months
  - ca. 1965
- Moore rev.2
  - 2x devices/chip every 24 [18] months
  - ca. 1975
- Self-fulfilling prophecy?
  - "it happened because everyone believed it was going to happen"
  - The receding brick wall

FIGURE 3 Moore's law.

Dr. Flamm pointed out that there was no actual physical basis underlying Moore's law; the process involved was that "human beings were investing in R&D and inventing new technologies." Moore's law was a well-informed prediction. Nonetheless, it became in essence a kind of coordinating device for an increasingly complex and dispersed industry. "Moore's law worked because everyone believed it was going to work," he said. "If you wanted to be competitive, you had to bring out the next generation of technology on schedule."

Another piece of legend, he said, was the history of "brick walls," technical problems or physical limitations that periodically threatened to slow or even halt the rapid progress of the semiconductor industry and derail Moore's law. Every time such a brick wall has been described, however, a solution has emerged, and the industry has continued to move ahead as predicted.

By the early 1980s, nearly 20 years after Moore's law was suggested, the United States no longer dominated the industry. In particular, Japan had invested a major effort in their own semiconductor technology, beginning in the 1950s. Some of their investments had paid off, especially in materials, equipment, and manufacturing technology. They were doing so well that many U.S. firms found themselves falling behind, especially in manufacturing technology, and customers of these firms began to discuss a perceived quality gap in U.S. chips. The industry initially denied it, but customers were doing their own tests on failure rates. In addition, the costs of U.S. firms were too high, and they were falling behind in manufacturing productivity as well.

By the mid-1980s, the industry was also hampered by a host of new trade issues and several national security concerns, including those identified by the National Science Board in 1986. These were based on the supposition during the Cold War that advanced technology, especially semiconductor content, constituted a large component of the nation's security and should be carefully guarded.

Evidence of Success

In this tense environment, the idea of a government-industry partnership arose as a mechanism to sort through these challenges and propose a unified strategy. Even though President Reagan traditionally opposed public interventions in the free market, the Republican administration took a favorable view of the partnership that became SEMATECH, as did the semiconductor firms themselves.

It is fair to say, according to Dr. Flamm, that SEMATECH became, under the aegis of the 1984 National Cooperative Research Act, the highest-profile government-industry R&D consortium in the United States. Despite that, however, there is a dearth of serious research on its impact. He said that the entire body of empirical literature amounted to three studies "with any pretense of rigor," none of them with quantitative empirical research that yields reliable proofs. That, he said, was why he was careful to acknowledge at the outset that his remarks were based on his own opinions, information gathered through interviews, and his reviews of published results.

The designers of SEMATECH took some time to plan a strategy, and they decided to emphasize strategies that would improve the equipment and materials used by U.S. producers. In 1982, William Spencer became the director of the consortium, and began to focus on reducing the time between new technology nodes and speeding up the flow of technology. He also promoted the sponsorship of a technology roadmap, which by now had become a fundamental feature of the consortium. The roadmap had been judged a "hugely" important and innovative model for global technology formation, adapted internationally. In the 1990s, the government subsidy ended, and SEMATECH continues today as an international organization.

Although his primary conclusion is "unprovable," said Dr. Flamm, he said that he found widespread agreement the partnership played a significant role in the resurgence of the U.S. semiconductor industry. There is little doubt that the industry did come back, he said, and that many U.S. firms today are on the leading edge of manufacturing—a condition that was not true in 1985. It is impossible, he said, to determine the extent to which SEMATECH was responsible. And there are critics in the industry today who ask to be left alone to chart their own course—although, says Dr. Flamm, this request was not heard in the mid-1980s.

Imitation as More Than Flattery

Perhaps a more interesting, and telling, consequence, he said, was that SEMATECH was widely credited in Japan with a considerable accomplishment. The proof of that came in the 1990s, when the Japanese began to copy the structure of SEMATECH in their semiconductor strategy. Another line of evidence, he said, had been the "revealed willingness" of the members of SEMATECH to increase their funding for the consortium to about $140 million a year after the government subsidy disappeared. All of these "data points," he suggested, showed that the effects of SEMATECH were widely viewed as useful.

Perhaps most importantly, said Dr. Flamm, the advance of productivity in the semiconductor industry accelerated dramatically in the late 1990s, after the major strategies of SEMATECH were installed. He cited Professor Jorgensen's view that this acceleration was directly linked to a simultaneous upsurge in U.S. productivity. "There was a direct link," he said, "between this improvement in the pace of introduction of semiconductor technology and the improvement of the aggregate macro-economic performance of the U.S. economy."

He returned to the evolution of the roadmap, which he said had "never occurred in any global high-tech industry before." It created the basis for an international framework to coordinate technology development on a global scale. Even more remarkable is that companies that are competing against one another voluntarily assemble every year to try jointly to identify technological obstacles and make plans to overcome them collectively as a global industry. "This is a new and totally unique phenomenon that I think is one of the lasting contributions of SEMATECH," he said "—a template for a new form of R&D collaboration."

He turned to a slide that illustrated the price decline of semiconductors in late 1990s, showing acceleration (see Figure 4). "This may not be entirely due to SEMATECH," he said, "but there certainly has been a pickup in the decline of semiconductor prices, and therefore an economic impact." He also showed an illustration of prices of memory and microprocessors, which also declined substantially. "Clearly something happened around mid-1990s," he said.

He concluded that SEMATECH was an interesting and successful experiment, a public-private partnership that operated on legal, financial, and technological levels. "It is widely believed by people who have some inside information to have been useful to the U.S. semiconductor manufacturers," he concluded, "and I think it has led to a unique new mechanism for international, industrial public-private R&D coordination. In many respects, the partnership is an institutional innovation that will live on long after SEMATECH itself it is no longer necessary."

Dr. Bement added that NIST had had a long partnership with SEMATECH, not only providing inputs for their roadmapping, but also taking NIST projects from the roadmap. In addition, SEMATECH annually critiqued the NIST program in areas related to semiconductors.

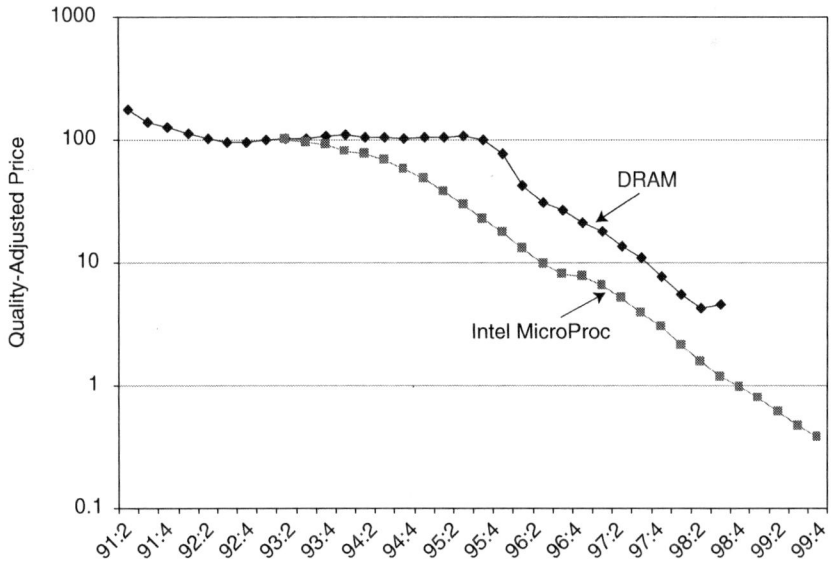

FIGURE 4 Change of pace?

PARTNERING FOR PROGRESS:
THE ADVANCED TECHNOLOGY PROGRAM

Maryann Feldman
Johns Hopkins University

Dr. Feldman presented the results of about 5 years' work as an independent consultant for the Economic Assessment Office at NIST, where she assembled a profile of facts and figures on the Advanced Technology Program. ATP, a public-private partnership program, had held 42 competitions that generated nearly 5,000 proposals and resulted in 602 "highly competitive" awards; less than 20 percent of the projects received funding (see Figure 5). An important component of the ATP project is that it involves partnerships among different types of organizations. Nearly 8,000 organizations participated in the ATP competitions, and 1,300 of them received funding and worked together on projects.

The ATP program provides an award to firms on the condition that they find matching funds. This is considered a critical aspect of the partnership. She reiterated Dr. Bement's observation that the ATP had funded about $270 million worth of R&D related to national defense. This translates, Dr. Feldman said, into a total investment in the U.S. economy of about half a billion dollars. Beyond that value, her work demonstrated that garnering an ATP award bestowed a "halo effect" that made it easier for participating firms to raise further funding. This created an even larger ripple effect in the U.S. economy.

- Forty-two Competitions: 1990 – June 2002
  - 4,969 Proposals received: 602 Awards
    - < 20% of projects receive funding
  - Partnerships among organizations
    - 7,985 Participants submitting proposals
    - 1,274 Participants in Awarded Projects
  - Leveraging Investment in R&D Projects
    - $1.9 billion in ATP Funds
    - $1.8 billion in Industry Contribution

FIGURE 5 Advanced Technology Program (ATP).

Some Causes of Market Failure

Dr. Feldman noted that as an economist she had been accustomed to believe that in most situations the market would naturally lead to an efficient allocation of resources. She learned, however, that there are several reasons why firms are likely to underinvest in R&D projects due to market failures that necessitate partnerships. The first—relevant to national defense—is the tendency of firms to avoid research projects that hold promise but have a high chance of failure. Firms prefer not to "get too far ahead of the pack."

A second reason for under-investment is that invention has become more complex, and many firms lack the in-house capacity to invest effectively in challenging R&D projects. R&D development often requires collaboration, but collaboration is difficult and costly. Therefore, firms have developed a bias toward short-term, go-at-it-alone projects and away from long-term projects that require multi-firm collaborations.

Finally, private incentives may not be sufficient to induce firms to undertake projects when they cannot be sure of appropriating the resulting benefits. This is a classic case of a market failure—when the private rate of return is lower than the public rate of return. The new knowledge or product is available freely to other firms and individuals in the economy, while the firm that created the knowledge or product is not able to price those benefits, or their knowledge spillovers.

These reasons lend support to those who advocate a government role in public-private partnerships—especially projects that promote pre-commercial technological development of high-potential social value. In order to assess the outcomes of the ATP program to do just that, Dr. Feldman undertook a survey of the 1998 ATP applicants. Prior to her survey, it was difficult to discern the net effect of the partnership mechanism. The ATP tracked the subsequent results of the award, but had no way of demonstrating what would have happened to those firms and their research projects in the absence of an award.

Therefore, Dr. Feldman's study followed all the ATP applicants for 1998—those who won awards and those who did not. The goal was to observe the differences in the two groups: their types of projects, partnerships, and behaviors. Thanks to the rigorous peer-review followed by ATP, the researchers could track the proposals on the basis of technical scores given by reviewers. This allowed them to see if the ATP selected the kinds of high-risk projects that could not proceed without a government subsidy.

The researchers followed the firms for one year after the competition, and then asked each firm if it had pursued its proposed project and if it had been able to secure additional or other sources of funding. The general conclusion, using regression analysis and controlling for firm characteristics and unique factors, was that the ATP was indeed selecting projects that had greater potential to generate substantial public benefits—primarily the riskier, early-stage projects and new partnerships. Also, the projects selected for funding had characteristics associated with knowledge spillovers and the propensity to share research results. It appeared that the ATP program managers examined the broader prospects of research proposals and did not simply select high-profile projects.

Debunking the Myth of "Agency Capture"

Dr. Feldman addressed the assumption that public-private partnerships are commonly involved in "agency capture." Agency capture occurs when the awarding agency returns repeatedly to the same successful companies, giving multiple awards. By applying controls to the evaluation, she found that agency capture was not in fact happening.

Dr. Feldman found that of the firms who applied and were not selected for funding, the majority of proposed projects—70 percent—did not proceed and were abandoned by the proposing firm; 30 percent did proceed, but at a reduced level. The firms that did proceed were those able to raise money outside the ATP competition. In fact, firms not funded by ATP were more likely to seek external funding. However, the firms that received awards seemed to develop a "halo," meaning these firms were able to attract three times the funding from venture capital firms and other private-market sources as the control group of firms with the same characteristics. Thus, Dr. Feldman's group concluded that the ATP program, instead of "crowding out" potentially worthy firms, as proposed in the literature, was actually "crowding in" more investment to the most worthy R&D projects.

Dr. Feldman further concluded that partnerships funded through ATP have direct relevance to national security concerns (see Figure 6). Two years ago, Johns Hopkins received funding from an anonymous donor for an information security institute, and Dr. Feldman was asked to be the policy director. Subsequently, she reviewed the most promising ways to inject new ideas into the security arena. She saw that ATP, because it relies on commercial firms to propose projects, created the ability for the private sector to clearly see and invest in the kinds of innovative applications that have highest potential to bring new capabilities to bear on national security concerns.

Beyond that, Dr. Feldman said her research results suggest that the ATP offers incentives that tend to increase the efficiency in the overall national system of innovation. The projects proposed to ATP are private-sector solutions—the kinds of high-risk, high-payoff creative ideas that are not likely to proceed without a public-private partnership. The ATP is grounded through a rigorous peer review. This grounding is reinforced by the participation of NIST, the parent agency of ATP, which sets standards for the infrastructure and platforms of national security.

Finally, Dr. Feldman concluded that her research results suggest that the ATP actually increases private-sector R&D in the kinds of activities that it funds. This means that ATP should be considered a program that provides a well-

- **Mid-IR Cavity Ring-down Spectroscopy** — BlueLeaf, Inc. (Sunnyvale, CA)
- **Multiplex DNA Diagnostic Assay Based on Microtransponders** — PharmaSeq, Inc. (Monmouth Junction, NJ)
- **Certifying Security in Electronic Commerce Components** — Cigital, Inc. (Dulles, VA)
- **A Master Patient Index (MPI) for Massively Distributed Records Across a U.S. National Backbone** — Sequoia Software (Columbia, MD)

FIGURE 6   ATP-funded projects with relevance to national security.

functioning infrastructure of research funding that can aid in finding creative platform solutions related to national security. She closed by pointing to the ATP's "great Web site," <www.atp.nist.gov> where visitors can review the technologies and companies that are likely to be relevant to improving national defense.

UNIVERSITY RESEARCH AND THE MARKET: THE CARNEGIE MELLON EXPERIENCE

Christina Gabriel
Carnegie Mellon University

Referring to the title of her talk, Dr. Gabriel said that to some people, "university research" and "market" did not fit well within the same title. In fact, she said, the two worlds can interact productively in ways that are relevant to partnering against terrorism. She had been impressed by how many academic people at her own institution were looking for ways, in the wake of September 11, to adapt their own research activities to the new and pressing needs of their nation. Just after that event, over 40 faculty members from across the campus attended a meeting convened hastily on a weekend, so that they could strategize with each other about ways they could use their combined expertise to help.

She said that Pittsburgh, where Carnegie Mellon is located, has a rich history of interaction between academia, industry, and the community at large. A strong entrepreneurial economy was created in Pittsburgh by Andrew Carnegie and other well-known industrialists of a century ago—resulting in the dominance of U.S. Steel, Westinghouse, Heinz, Alcoa, and other large manufacturing companies (see Figure 7). Carnegie endowed his new university in 1900 so that it could educate the children of the steelworkers, and the University of Pittsburgh's gothic skyscraper was designed to be visible from all the working-class communities in the region as an inspiration to them. Then, in the 1970s, when Japan successfully implemented new, lower-cost techniques of steelmaking, Pittsburgh lost most of its steel jobs along with the mills themselves, and it has taken decades to restructure the economy around a more diverse set of industry sectors. Today, the region is looking to its universities and medical research centers as the source of new jobs and new companies to bring back rapid rates of growth.

One benefit Pittsburgh continues to enjoy from the accumulation of wealth by those early entrepreneurs, Dr. Gabriel said, is the availability of more philanthropic foundation dollars per capita than perhaps any other community in the country. In recent years, the Pittsburgh foundations have supported a variety of new programs designed to help the community capitalize on the quality of its universities to revitalize the economy. Connecting university research more closely with the market where jobs can be created is a goal Pittsburgh shares with many regions around the country. With the pace of innovation increasing, research in many fields actually is much closer to the market than it has ever been before. "Because of information technology," she said, "the pace of change is relentless. If I have an idea and I tell you about it, you'll probably have a better idea soon. Through the Internet, I can tell millions of people all at once—and then getting the next better idea is a race among those millions—very likely faster than either you or me." She recalled talking to a friend, Robert Colwell, head of the design team for the Intel Pentium II processor, who described the enormous efforts made by a very large team of people to make that chip the most powerful product on the market. After all that effort, he said, the chip would probably be obsolete after no more than a few years. Had he been an engineer in Caesar's day, he would have worked on the aqueducts or the Appian Way, which remained in use for centuries. But even though our results may be shorter-lived, the intellectual challenges are more exciting than ever. If we are to be more innovative, we need to become better at working within partnerships of various kinds.

The Importance of a Government Partner

As part of her research and management positions in industry, government, and universities, Dr. Gabriel has observed the characteristics of a variety of effective partnerships. She found that government funding played an important role, especially in the early growth stages of technology companies, "more often than people realize. Government is not just something that sits on the sidelines. Government does play in important role, especially as an early catalyst and in shaping

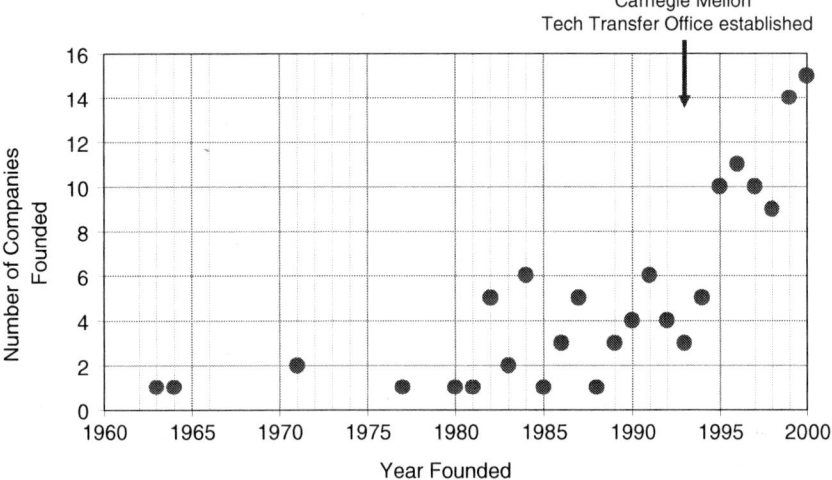

FIGURE 7   Estimated number of new Pittsburgh-region companies started with Carnegie Mellon founders or technology.

incentives in the marketplace, and we have to make that role as effective as it can be."

She proposed a list of factors that made university-industry partnerships successful. The most important, she said, was strong interpersonal relationships. "It's really about trust; about people knowing each other well enough to take risks with them." No matter how well one structures a new program at the policy level, she said, it is "extremely important to make sure the incentives make it work at the operational level" as well.

As an example of a government-supported partnership that led to regional economic development, she cited the NSF Engineering Research Centers. This program was designed to improve engineering education by encouraging stronger working partnerships between universities and industry. One ERC at Carnegie Mellon, the Data Storage Systems Center, partnered with a consortium of companies including Seagate, the largest disk-drive manufacturer in the world. In 1998, Seagate decided to create its first and only research laboratory. Although the company headquarters is in California, the company chose Pittsburgh for its new facility so that the collaboration with Carnegie Mellon's center could continue. For its first several years, the new lab grew at a rapid pace, hiring on average one PhD per week from over 20 different countries. Collaboration with the university continues to be strong, and Pittsburgh is developing as a hub for the information storage industry, with university spin-off companies also being created as a result of the ERC. Most of the economic activity has been supported by the private

> "Technology transfer" describes the movement of ideas, tools, and people among institutions of higher learning, the commercial sector, and the public.
>
> The **1980 Bayh-Dole Act** gave intellectual property rights to organizations that perform research with federal funding, as an incentive for commercialization of federally funded inventions.
>
> Before Bayh-Dole: <250 U.S. university patents per year
>
> Today: about 1500 U.S. university patents per year, 2000+ companies spun out of universities
>
> **Now that 20 years have passed....**
> **What have we learned? How can we do better?**

FIGURE 8 Technology transfer.

sector, but the early federal dollars invested in this peer-reviewed center award was the catalyst that made this growth possible.

The Learning Curve of Tech Transfer

To stimulate the contributions that Carnegie Mellon technologies can make to economic growth, the university has restructured its technology transfer function. Starting in late 2000, Dr. Gabriel led an exercise, marking the 20th anniversary of the Bayh-Dole Act, that brought together people from across the university "to see how we can do technology transfer better."[9] Before Bayh-Dole, fewer than 250 U.S. university patents were issued each year; today, about 1,500 U.S. university patents are issued each year, and thousands of companies and jobs have been created that are based on university research (see Figure 8).

She said that one result of this increased patenting activity by universities is anger and frustration on the part of many of the private-sector constituencies that the university interacts with. Whether as nonprofit organizations that receive federal funding or as public institutions, universities operate under a set of legal constraints that are not well understood by outsiders. It can seem to industry

---

[9]The Bayh-Dole Act of 1980 (35 USC 200-212), along with its later amendments, was highly influential in stimulating academic research by allowing universities and academic researchers to patent and benefit financially from the results of government-funded research.

partners as well as to faculty inventors that the university is overly bureaucratic. People from the business world, investment world, and academic world often seem to speak different languages.

"How do we fix that?" Dr. Gabriel asked. She noted that a national organization, the Association of University Technology Managers (AUTM), exists to help universities share best practices and promote the movement of innovative ideas into the private sector. Each year AUTM compiles data collected from about 200 universities and research institutes to measure the success of technology transfer, such as how many patents are filed, how much royalty revenue is realized, and how much money is spent on legal costs. However, there is a shortcoming of such financial metrics alone, she said. They say little about social return—how easily new inventions move into broader use, benefits to the region around a university, improvements in faculty attraction and retention, etc. If the sole focus of technology transfer is on financial returns to the university, she argued, it is difficult to make the case for the university to support a tech transfer mechanism at all. The reality is that a technology transfer office typically costs more to operate on an annual basis than the university realizes in ongoing royalty revenue. The function must be subsidized virtually every year; only occasionally, with luck, does a university spin off a useful product that brings in millions of dollars in revenue—and even then, the return comes years after the investment of staff time and legal expenses. There are too few of these "home-run hits," she said, to justify tech transfer offices for most universities on purely financial grounds (see Figure 9).

Commercialization
3687 licenses and options
344 company starts
417 new products
Sales support 270,900 jobs

Intellectual Assets
5339 new patent applications
1 per $4.8 million

Discovery
11,607 disclosures
1 per $2.2 million

Research
$25.7 billion

Gross proceeds to a university from commercialization are typically no more than 1-3% of sponsored research revenue — except for the "lucky hits"

FIGURE 9 U.S. university tech transfer. SOURCE: Jim Severson, from Association of University Technology Managers (AUTM) national survey data, FY1999.

Contrary to the dreams of many universities in the early days following Bayh-Dole, tech transfer can be expected to bring in only a small percentage of what the university earns for its federally funded research.

To test that assumption, she plotted technology transfer revenue for one recent year at the largest of the nation's universities, divided by overall research revenue (see Figure 10). To avoid absolute-dollar comparisons between the giant California system, for example ($1.6 billion of research revenue), and smaller schools like CMU, this ratio roughly scales the universities by size. In general, universities did well if they realized a few percent of the research budget as technology transfer revenue. For a $100 million research budget, bringing in $3 million in royalties and capital gains revenue "should be considered a very good result," she said. "But don't hope or expect that you'll do even that well routinely."

During Carnegie Mellon's reassessment of its tech transfer function, the university group first asked a large number of experienced people what they thought was wrong with the process and how it ought to work instead. Tech transfer was being regarded as a bureaucratic function that was added to the university, rather than an integral part of the institution's mission. The philosophy in the early years was more or less to search for the "diamond in the rough," focusing all its energy on that "best bet" that might hit the jackpot for the university—while providing little or no attention for ideas that seemed less promising financially. The implicit goal was to maximize revenue while minimizing risk. "But those two goals are

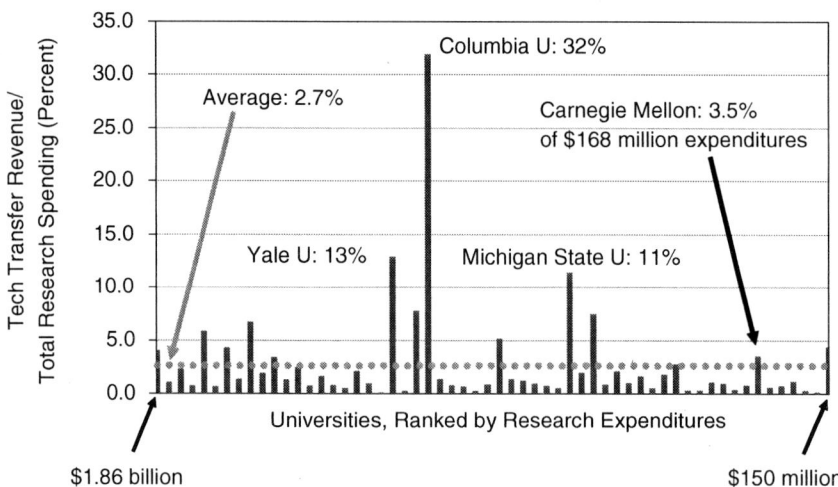

FIGURE 10 FY 1999 tech transfer revenue as a fraction of total research spending, top spending universities.

mutually exclusive," she said. "There's no way you can have zero risk and maximum return. It wasn't working." Moreover, since even the best venture capitalists, who are far downstream in the commercialization process, get "home runs" less than 10 percent of the time, the group felt that a university should aim for a higher volume of transactions rather than trying to make these guesses at the research stage.

Debunking the "Home-Run Strategy"

She backed up to say that on one level, it was possible to argue that the strategy was working beautifully: The University had had an income bonanza in 1998 when the Lycos search engine was invented at Carnegie Mellon by Fuzzy Malden and his collaborators (see Figure 11). The university had equity in the company, and when it went public, the university's half of the capital gains income provided enough money to build a much-needed building, while the other half went back to the inventors. The "home-run strategy" worked in that case, and it was possible to argue in hindsight that the result was good for everybody.

But the Carnegie Mellon task force decided on balance to change the strategy. While Pittsburgh was not the booming economy of Silicon Valley or Boston, it had by then gathered a substantial population of experienced lawyers, accoun-

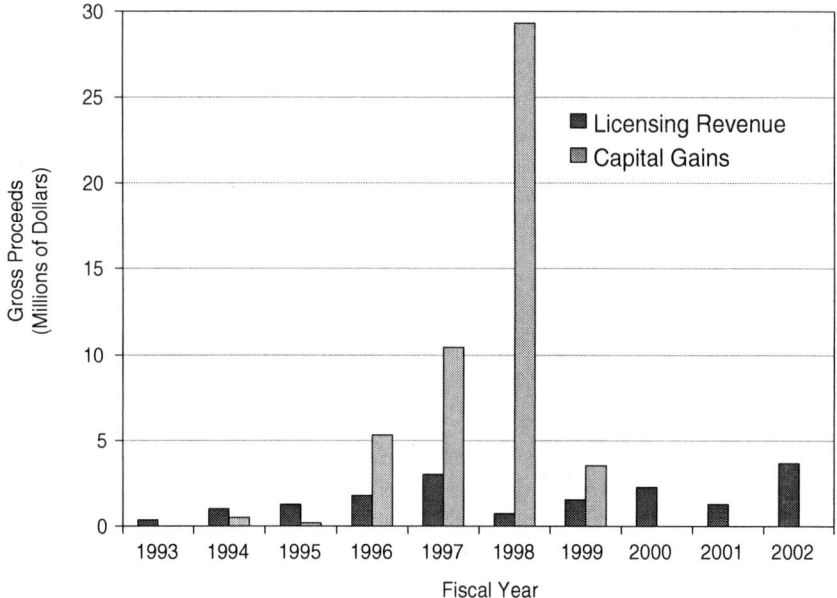

FIGURE 11 Carnegie Mellon tech transfer revenue.

tants, and economic development organizations that could offer competent help for new entrepreneurs. With more expertise in the community, tech transfer could now more effectively be done in close partnership with the external business community. So the university decided to simplify and streamline the transfer process, making it easier for an embryonic company to get out of the university quickly. After that, the company would depend on community resources, with only informal help from the university and its network of spin-off companies.

The Power of Group Process

Perhaps the most important shift in tech transfer design was suggested by the influence that Nobel Laureate Herbert A. Simon has had on Carnegie Mellon. Simon, said Dr. Gabriel, "really understood how to have strong people in different disciplines work together to do amazing things." Carnegie Mellon is now known for its collaborative, problem-solving culture across the campus, from the fine arts to engineering, computer science, and business. The task force decided to try to apply that gift to their process of innovation transfer. Evaluation of new technology concepts is now done through a collaborative, real-time evaluation of the innovation by a panel of reviewers with complementary expertise. Half or more of the reviewers are chosen from the business community outside the university in order to involve deep experience with commercialization relevant to

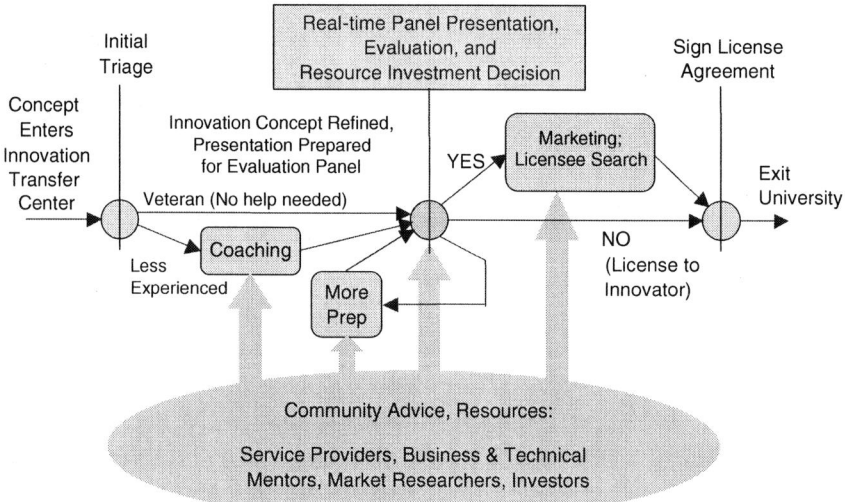

FIGURE 12 Innovation transfer process.

the innovation to be considered. The group has a conversation about commercialization strategy around a table with people they may not yet have met, who share their interest in the technology but see it from a different perspective. Through this brainstorming session, the university gets high-quality input to help make better decisions about its resource investments. The reviewers get an intellectually stimulating experience and the chance to expand their own personal networks. The creators of the innovation get higher-quality attention and a faster decision from the university about how they should proceed. By making judicious use of appropriate external expertise, the university makes these difficult decisions and carrying out these tasks easier and more likely to be successful.

This interactive approach is designed to bring out issues that might be missed by individual specialist reviewers and to offer a range of perspectives for the university to consider. In addition, since each panel will involve different experts, many of them from the Pittsburgh region, the process will, over time, strengthen the network of technology and business professionals in Pittsburgh and the working relationship between Carnegie Mellon and the regional community. "At the end of a two-hour session," she said, "you know everyone around the table well enough to ask them anything, anytime. The network of experts that this builds can really help our inventors and their spin-off companies."

She then discussed how to extend these discoveries about group interaction to a more effective response to the national focus on preventing terrorism. Like others at the workshop, she was optimistic about the ability of public-private partnerships to provide effective structures and to speed up the creation and deployment of useful new innovations. She based this opinion partly on her own experience with the Technology Reinvestment Project[10] during her tenure at the National Science Foundation in the early 1990s.

The TRP, she recalled, was politically charged from its first implementation, dividing those who saw an urgent need to use private-sector advances and the "dual-use" concept to strengthen the defense industry sector and those who believed the program was primarily trying to move spending away from defense to commercially directed applications. As the program began, she said, one of the greatest needs was to break down the walls that divided different program offices in different government agencies funding work in similar technology areas. "When I arrived at NSF in 1991," she said, "I was appalled to find that program

---

[10]"The Technology Reinvestment Project was designed under the first Bush administration and implemented early in the Clinton administration as an interagency partnership to help implement its 'defense reinvestment strategy.' The TRP was administered overall by the Defense Technology Conversion Council (DTCC) and chaired by the Defense Department's Advanced Research Projects Agency (ARPA, now DARPA). In 1993–95, six federal agencies reviewed TRP proposals together; all funding recommendations were also made jointly."

officers in my technology field did not even know their counterparts in other agencies—even though we were all working within a few miles of one another."

The cause of the "walls" was not mysterious, she observed: Each agency had its own mission and everyone was busy. What the TRP did was "to force six agencies with a common pot of money and an important overarching objective to jointly structure the program, review proposals, make decisions, and execute tasks." The agency representatives deliberated together in intense review sessions for weeks, driven by a requirement that all the money had to be disbursed in a finite time. After that experience, she found that those six people knew and respected each other very well. "We all remember how exciting and important that was. I truly believe that if we did joint program execution like this more often, the government would work better. Business-as-usual has too often been for each agency to create its own new program for each new hot field, not communicating with the others and ignoring themes that require a larger critical mass than any one program can provide. It would be nice to know that whenever there's an urgent national need, you could get an interagency group like this together to find and fund the best collaborative work quickly, without turf getting in the way."

She closed by noting the value of other major federal-private partnership programs designed to stimulate technological innovation and economic growth, such as NIST's Advanced Technology Program and the Small Business Innovation Research (SBIR) program.[11] The SBIR set-aside program, for example, she said, has helped a large number of new technology companies create and commercialize innovations much more quickly than could have been done within large companies. Much of this work is done in partnership with universities, and much more could be done this way. The SBIR mechanism could play a major role in seeking out and developing innovations to help the nation address its new technology challenges within the war on terrorism. She suggested that program organizers study the feasibility of adapting SBIR to this need, especially by finding ways to reduce the cycle time for proposal solicitation and review. This, she said, might effectively tap the current energy in universities and companies and make their work quickly responsive to national needs.

DISCUSSANT

Michael Borrus
The Petkevich Group

Mr. Borrus said he had been struck by the parallel between the qualities of a good public-private partnership and those of a long marriage. Each had to navigate

---

[11]See, for example, National Research Council, *The Small Business Innovation Research Program: Challenges and Opportunities*, Washington, D.C.: National Academy Press, 1999.

the uncertainties of misunderstandings and poor communication, but both had the potential for high achievement over the long term. And when they did fulfill their potential, they could accomplish goals that were beyond the reach of a single person. In the same manner, public-private partnerships can advance the pace of technology in ways that neither federal agencies nor private firms alone can do: by overcoming roadblocks that block development and innovation; by focusing creative thinking on national needs that would otherwise go unmet; by compensating for market gaps or imperfections; and by producing significant social benefits that an individual firm or sector would be unlikely to achieve.

He noted that the United States had had a long history of public-private collaborations, dating back perhaps to Alexander Hamilton's *Report on Manufactures*, and certainly as far as the Morrell Act of 1862, when the government established the land-grant universities and the agricultural extension service to assist private farmers. More recently, partnerships played a significant role in developing such innovations as the jet air frame and jet engine, the transistor and the silicon chip, computer technology and the Internet, and many vaccines and other elements of modern biotechnology. "It's worth reminding ourselves," he said, "that if we do this right, these collaborations can work, and we can apply the lessons learned to the extraordinary goal of homeland security."

Attributes of Successful Partnerships

He said that partnerships are particularly effective at tackling very complex problems, especially those that occur at the intersection of existing disciplines, methodologies, and perspectives. He suggested some key attributes of successful partnerships:

- He agreed with Dr. Gabriel that trust plays a critical role in bringing together people with different knowledge.
- The initial design of partnerships also appeared to matter, he said. One size does not fit all, and each partnership must be designed according to its specific goals. This does not mean, however, that program planners should attempt to pre-determine how to reach those goals. In fact, he said, failure is almost always the result when a group attempts to over-plan.
- Competition is important throughout the process, he said, even among members of the partnership itself.
- Those projects that succeed tend to have funding commensurate with their goals; either too much or too little funding can impede progress.
- All members of a partnership should play the roles for which they are best suited, rather than assuming or being asked to play a role for which they are not institutionally or historically prepared. For example, universities are not naturally suited to play the role of venture capital firms; they are better able to facilitate interaction between the VC industry and university researchers.

- Partnerships should receive frequent, rigorous evaluation throughout their lifetime. Evaluators must have the courage and willingness to point to experimental failure or error. For some partnerships that have failed over the years—including the original Synfuels projects, the fast breeder reactor, and the supersonic transport airplane—evidence of experimental failure was sometimes ignored and yet the program was permitted to continue—and to fail. External input may force the evaluation toward greater objectivity; inbreeding can be fatal to sound assessment.[12]
- Partnerships must be flexible. He said that virtually no venture capital company had ever funded a company that exactly followed the business plan originally used to secure funding. The key to a successful venture, he said, is to support people who can adjust flexibly to new markets, new technologies, and unexpected experimental results.

The Need for Validation by the Market

A subtext to all public-private ventures, he said, was the need for validation by the market. There must be substantial overlap between what the project is trying to accomplish and where the relevant civilian commercial markets are headed. In past projects that failed, such as the SST, there was substantial divergence between partnership objectives (smaller planes, supersonic speeds) and what the market wanted (wide-bodies and long-haul). As technology development proceeds over time, it becomes ever more difficult and expensive to force the two sets of objectives together if they've diverged from the start. Reaping the full social and economic benefits of collaborations requires sensitivity to commercial market demand on the part of all partners.

The fight against bioterrorism in particular, he suggested, heralds a profound shift in U.S. defense. He cited George Poste, former chair of research and development at SmithKline Beecham, who predicted within the next several decades a

---

[12]All three of these programs were costly and controversial, and fell short of commercial success:

- In the 1970s, a consortium of energy companies obtained federally guaranteed loans to finance the construction of the Great Plains Synfuels Plant. Operations began in 1984, but the consortium abandoned the plant in 1985. DOE assumed ownership in 1986, and in 1988 sold the plant to a private company. Today the plant produces synthetic methane gas and fertilizer from lignite coal.
- Fast breeder reactors use plutonium as a fuel and produce ("breed") more of it during operation. Development in the United States began in the 1950s, but was effectively ended when Congress stopped funding for the Clinch River fast breeder program in 1983. France, Japan, Germany, and the UK all experimented with fast breeders.
- The dream of an American-produced supersonic transport (SST) ended on March 24, 1971, when Congress voted to end funding. Over $1 billion was spent on development, but no planes were built. A French-British consortium did develop a supersonic transport, but only 16 Concordes were built. Development costs were never recovered, although the planes were operated by British Airways and Air France until 2003.

time in which every cell system in the human body will be understood sufficiently to be manipulated either for good—to cure disease—or for ill—to cause harm. Consequently, said Mr. Borrus, the nation faces no choice but to develop something new—a "true biodefense industry."

In doing so, he said, we are on the verge of attempting something not done since the post-World War II years: to coordinate public spending, academic research input, research from the national labs, and the activities of private industry to build defense capability. Today this capability will be in biodefense; 50 years ago, it was in conventional and nuclear deterrents. At that time, we made the choice to prevail on the basis of superior technology rather than superior numbers. And today, he said, we have the same choice, and we need to make the same decision. "We need to differentiate our performance in the biodefense realm on the basis of our superior technology, not in any other way."

He went on to demonstrate that partnerships were essential in developing our defense 50 years ago, and should be so again in developing biological capabilities.

Again, 50 years ago, partnerships brought tremendous technology-transfer benefits to civilian society, in addition to building a superior military capability. Today, he said, partnerships in biology have the potential to deliver benefits of the same order. To detect a biothreat, for example, one needs early warning, preferably at a point of care, so intervention can begin immediately. One also needs early intervention to contain the threat before it can spread. He suggested that such early detection and diagnosis are precisely what civilian society needs to maintain a strong health care system for the United States.

"If we do this right," he concluded, "by learning lessons from past public-private partnerships and applying them systematically to build the best biodefense capability, we also have the hope of improving our civilian health system and generating commercial leadership across a range of new technologies. We should consider the presentations today, and the panel, to be a modest but essential contribution to getting it right."

DISCUSSION

Anne Solomon, of the Center for Strategic and International Studies in Washington, commented that the relationship between partnerships and commercial markets was complex, and wondered what kinds of markets there would be for some of the technologies produced. She suggested two possible sources: government acquisition, and the commercial market, including sectors that control technologies, or companies that may want the products. She said that the CSIS had been developing a long-term strategy for bioterrorism countermeasures. The problems they found were enormous, she said, especially those having to do with the market and with the government's ability to articulate what it wants; a reluctance to talk about vulnerabilities for fear of liability; and competition, in the sense that one company hesitates to invest in protection unless its competitors also invest. How, she asked, would venture capital firms react to a company that invested in R&D that was aimed at counter-bioterrorism products if there was no known commercial market?

Mr. Borrus responded that if the potential reward was commensurate with the risk that needed to be taken, it would be taken. He noted as an aside that liability issues per se had not restrained financial misconduct at Enron Corp. because the potential for gain was so high.

A participant observed that the government may have to play a lead role in stimulating a market for biodefense products. This would not necessarily be bad, she said, especially since "companies almost always misjudge the market in terms of timing."

Dr. Bement added that the issue of product liability was central to the mission of NIST in the sense that standards and measurements are fundamental to improving reliability and reducing risk. "The whole infrastructure in standards development," he said, "is pretty much aimed at reducing product liability." Also, he said, even though public sector intervention is normally intended to address "market failure," this is a very subjective term. "It's in the eyes of the beholder," he said, "and something we don't know how to deal with very well."

Dr. Wessner pressed Dr. Flamm about the evidence that SEMATECH was regarded positively—that it had maintained its membership and added international membership, that companies had continued to pay their dues, and that within the industry it was considered to have a positive impact. He asked whether there was any additional "proof of success," and whether SEMATECH had been the model for the similar European consortia.

The Difficulty of Proving Successes

Dr. Flamm replied that it is difficult to prove success "when the target is so broad"—that is, restoring the competitiveness of manufacturing processes within U.S. companies. So many forces contribute, he said, that the idea of isolating a single causative factor is "basically impossible," unless one can actually measure inputs and outputs, which is "relatively hard to do for an entire industry." He argued that the only useful yardstick is the perceptions of the people involved: Those who provided the funding perceived that the program was worth supporting. In addition, Japan had explicitly modeled its program on SEMATECH. "There's great irony there," he added, "because to some degree SEMATECH was a response to Japanese programs of the 1980s." Today, he concluded, most industrialized countries now have some program resembling SEMATECH.

Mr. Borrus agreed, adding that the strongest indication that SEMATECH added value to the industry was the willingness of industry to fund it after government support ended.

David Peyton, of the National Association of Manufacturers, posed a question for Dr. Gabriel. Referring to recent testimony before the Senate Commerce Committee on nanotechnology, he described the opinion of a senior scientist from Hewlett-Packard that U.S. universities had become more difficult to partner with on sponsored research, because they insisted on tougher policies on intellectual property rights. He asked whether in fact leading research universities were trying harder to retain patent ownership, and whether it would be possible to develop a balance that would allow universities to spawn new businesses and major R&D companies to benefit from research done at universities.

Adjusting to the Age of Intellectual Property

Dr. Gabriel said that universities were indeed becoming more savvy than they had been in the early 1980s, and that IP was "a tough domain." Companies had once approached universities, she said, with the expectation that universities knew little about IP issues. As universities entered this domain with IP offices of their own, they had been "stumbling sometimes" as they moved along the learning curve to navigate the constraints of nonprofit law, IRS rulings, export controls, and the preservation of an open research environment. But she pointed out that companies also find each other hard to deal with, because intellectual property issues are "inherently contentious." "This is just a fact of life," she said, "and we're all going to have to learn from each other about how to make it easier."

Fred Adler, of WDC USA World-Wide, who identified himself as a former colleague of Dr. Flamm at the Department of Defense, said that he had been in Tokyo during the Sarin nerve gas attack on subway passengers. He asked whether it would be profitable to partner with the Japanese, in the assumption that the Sarin attack had served as a "technology accelerator." He also asked how public-private partnerships might be arranged with Japan, especially in the context of the planned global disaster information network.

Dr. Flamm agreed that there are strong anti-terrorism resources outside the United States, especially in Europe and Israel, and that he hoped we are tapping these resources. Mr. Adler added that there is good technology in many places, but that few of them have sufficient funding for investment.

Mr. Borrus said that "we have no choice but to partner; knowledge is too widespread around the world for us to do it ourselves." He noted that patterns of direct foreign investment already show that private companies know this and are exploiting technical specialization around the globe. He urged efforts to locate partnerships in the U.S. whenever possible so as to gain the most value from knowledge spillovers.

Dr. Bement closed the discussion by observing that NIST partnered with about 40 national metrology institutes around the world, bringing many of their members to NIST as guest scientists. He found that some of them, especially those from Japan, contributed valuable insights through their strengths in analytical chemistry, detection technology, and other fields.

Panel III ———————————————————

Partnerships Against Bioterrorism

INTRODUCTION

Larry Kerr
Department of Homeland Security

Dr. Kerr observed that this panel would present "the absolute experts in this area" who had gained from "nearly a year's worth of hindsight on the events of last fall." He recalled that "several distinguished panels" had already met over the past year to discuss the strengths and weaknesses of our bioterrorism preparedness. These include the NRC's Branscomb and Klausner report,[13] a blue-ribbon panel of the National Institute of Allergy and Infectious Diseases, the Presidents' Council of Advisors on Science and Technology (PCAST) letter, the President's national strategy on homeland security, and the House and Senate draft legislation for the new Department of Homeland Security.

All these documents had a common theme, he said: our strongest weapon against terrorism will be partnership. And in partnering, we have to create a technological and operational advantage over enemies. "As we work to create the Department of Homeland Security," he said, "the transition planning office has been mandated by the President and by Governor Ridge to create a

---

[13]National Research Council, *Making the Nation Safer: The Role of Science and Technology in Countering Terrorism*, op. cit.

groundswell within the nation's R&D efforts and to establish a new paradigm for preparedness."

He said that his office had been mandated to find ways to bring disparate partners together into mutually beneficial collaborations: government with academia, government with the private sector; federal with state and local partners; big business with small business; military with civilian.

The office planned to bring scientists, engineers, intelligence, and law enforcement groups together to focus not only on research and development, but also on testing, validation, and evaluation; on procurement and distribution; on concepts of operation; on all vital technologies necessary to defend the homeland. He noted that the office had a unique mandate within the bio-defense arena, and a unique opportunity. "I am constantly reminded that as we seek to guard our people and agriculture against biological attack," he said, "we do so by strengthening the resources and technologies that ultimately contribute to our public health infrastructure." An improved arsenal of diagnostics, therapeutics, and vaccines; better protective equipment for first responders; and a national disease surveillance system—all help improve the public health while offering powerful deterrents to biological attack. "This National Academy of Sciences symposium," he said, "could not have come at a better time for us."

PARTNERING FOR VACCINES: THE NIAID PERSPECTIVE

Carole Heilman
National Institute of Allergy and Infectious Diseases

Dr. Heilman began by saying she would offer some perspective on the National Institutes of Health (NIH), especially on their vaccine programs, and then focus more specifically on NIAID's biodefense activities. She would refer often to partnerships, she said, because "one of our assets at NIH is our history of partnering."

NIH supports several but not all aspects of the vaccine R&D pipeline, she said, and this necessitates partnerships with other entities that bring needed skills to vaccine development. This development process is long and tedious, she said, both because vaccines are difficult biologics and because regulatory hurdles slow the process. Thus, partnerships make both technical and financial sense. About 90 percent of NIH's budget goes off-campus to support the activities of academic and business researchers; the balance covers intramural research programs and support functions at NIH.

Vaccine development requires information that is difficult to produce before decisions can be made, she said. Researchers have to understand the pathogen, its components, and how it interacts with the body. Only when those questions are answered, can researchers begin identifying targets and move from those targets toward making the tools that will be needed for actual development of a vaccine.

Once a vaccine is made, laborious preclinical testing begins. This is often a cyclical process—an initial try, adjustments, a retry, tweaking again, retrying, and so on. After that preclinical work, a sample lot must be produced and approved for human trials—a "huge hurdle." The vaccine must go through a series of clinical tests to assure that it is not only effective but also extremely safe. Because vaccines are generally given to healthy people, the standards required for ensuring safety are very high.

This process requires multiple players, all of whom contribute in different ways to new or improved vaccines. The NIH in general conducts basic research and develops medical interventions. The Centers for Disease Control and Prevention (CDC) is involved in surveillance, training local response teams, and maintaining stockpiles of vaccines and antimicrobials. The Food and Drug Administration (FDA) is responsible for regulating vaccines, therapeutics, and diagnostics. Both large and small companies contribute in different ways toward vaccine development. In addition to the academic community, there are strong partners among nonprofits and global organizations. The Gates Foundation, in particular, provides critical "pull," guaranteeing purchase of vaccines for developing countries. This provides a needed incentive to many companies.

Partnering for Vaccine Development

The NIH has long maintained robust vaccine development programs, she said, and they have become more robust as a result of post-9/11 activities. Virtually all of the vaccine development of NIH involves partnering in a variety of ways. This requires flexibility on the part of both partners, especially for industrial partnerships. Each partner needs something different, or they need to negotiate in a different way.

She turned to biodefense, and why the Department of Health and Human Services (DHHS, the parent agency of the NIH) is concerned about vaccines as they relate to this topic. Traditionally this has been the domain of the Department of Defense (DoD), but for three reasons the NIH has now moved into this area.

First, biodefense is different from biowarfare. Vaccines developed by the DoD are usually developed with the goal of preparing troops for situations where they may encounter a bioweapon. The licensed anthrax vaccine, which involves six shots over 18 months followed by yearly boosters, can be used for troops, but is not helpful in the event of a terrorist attack. Because it is impossible to guarantee an amount of time to prepare for a bioterrorist attack, the goal is to develop a vaccine that works quickly.

Second, in the military, the population to be defended is fairly homogeneous. The population at large is not homogeneous; it includes infants, people with HIV, and people on chemotherapy, the well elderly, and other special populations with less than robust levels of immunogenicity. This requires different approaches to prevention and treatment.

Third, the threats one may encounter in civilian settings are usually different from those in military settings. A terrorist might decide to lace all muffins in a coffee shop with giardia and succeed in causing considerable confusions and even panic; in a military situation, soldiers with giardia would be expected to tough it out.

She said that the two most important pathogens on the list of category A pathogens are smallpox and anthrax; the third is Ebola virus (see Figure 13). Category B and C priority pathogens contain a larger number of biological agents; little is known about some of these pathogens.

A Six-Fold Increase in Funding for Biodefense

The Department of Homeland Security, in a document about strengthening the nation, recognized the importance of vaccines, antimicrobials, and a strong infrastructure for medical research. As a result of many factors, including that publication, said Dr. Heilman, the biodefense budget of the NIH was increased about six-fold from FY2002 to FY2003 (see Figure 14).

Little is known about some of the potential agents of bioterror, because of the difficulty of studying them in either a natural or laboratory setting. Consequently, some of the additional funds are being used to develop a robust basic research program, including genomics and proteomics. More Biosafety Laboratory (BSL)-3 and BSL-4 facilities are being built to address the safety of individuals working on pathogens and the general public. One of the immediate

Biological Agent(s)	Disease
Variola major	Smallpox
Bacillus anthracis	Anthrax
Yersinia pestis	Plague
Clostridium botulinum (botulinum toxins)	Botulism
Francisella tularensis	Tularemia
Filoviruses and Arenaviruses (e.g., *Ebola virus, Lassa virus*)	**Viral hemorrhagic fevers**

FIGURE 13 Biodefense: Category A agents. SOURCE: CDC.

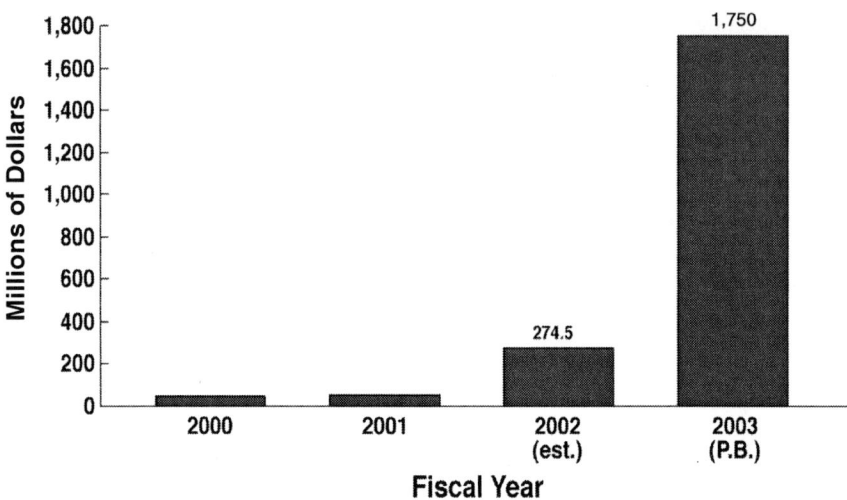

FIGURE 14 NIH biodefense research funding, FY 2000–2003.

- ■ Median age: 59 years (range 43 to 94)
- ■ 7 men, 4 women
- ■ Occupation:
    - postal workers: 6
    - mail handlers or sorters: 2
    - journalist: 1
    - hospital supply room worker: 1
    - retiree: 1
- ■ Survival of patients (55%) markedly higher than previously reported (<15%)
    - rapid diagnosis
    - aggressive therapy with multi-drug antibiotic regimens
    - state-of-the-art supportive care

FIGURE 15 Eleven cases of bioterrorism-related inhalational anthrax in the United States, 2001.

goals is to distribute research finding to scientists through bioinformatics resource centers with databases that will allow scientists to access a large amount of genomic and related data.

She added that the American people expect products at the end of all this research, so there are specific objectives focused on producing drugs, vaccines, and diagnostics. In addition, the agency plans to expand the clinical research component, which must accompany the development of any particular product.

The NIH has an understanding with Congress that these projects require a long-term commitment of support in order to succeed. Providing homeland security is a continuing need and objective that cannot be reached in a brief burst of activity, no matter how well funded. As the need for research facilities is gradually met, researchers will continue to need flexibility to develop drugs and vaccines, and conduct basic research that makes them possible.

It is an NIH tradition to recruit groups of outside experts to provide objective guidance and help develop appropriate strategies for addressing timely issues. In February 2002, she said, NIAID convened a panel of experts to develop a strategic plan to guide the implementation of basic and translational biodefense research emphasizing specifically the Category A priority pathogens (see Figure 16). At the end of October, the institute repeated this exercise for Category B and

Biological Agent(s)	Disease
Category B	
Coxiella burnetii	Q fever
Brucella spp.	Brucellosis
Burkholderia mallei	Glanders
Burkholderia pseudomallei	Melioidosis
Alphaviruses (VEE, EEE, WEE)	Encephalitis
Rickettsia prowazekii	Typhus fever
Toxins (e.g. Ricin, Staph. enterotoxin B)	Toxic syndromes
Chlamydia psittaci	Psittacosis
Food safety threats (e.g. *Salmonella spp.*, *E. coli* O157:H7)	
Water safety threats (e.g. *Vibrio cholerae*, *Cryptosporidium parvum*)	
Category C	
Emerging threat agents (e.g. Nipah virus, hantavirus)	

FIGURE 16   Bioterrorism: Category B and C agents. SOURCE: Rotz et al., *Emerging Infectious Diseases*, February 2002.

C priority pathogens. From those meetings, the institute determined that it should be prepared to cover a wide variety of areas relating to research on the biology of the microbe, host response, and basic and applied research aimed at developing diagnostics, therapeutics, and vaccines against these agents. These areas include genomics, proteomics, antimicrobials, vaccines, and the expansion of research capability. The results of the meetings guided the institute in building its budgetary plan.

Goals for Biodefense at NIAID

The groups set the following goals:

- Conduct basic research on the biology of the microbe and host response;
- Conduct basic and applied research aimed at developing diagnostics, therapeutics, and vaccines against these agents;
- Develop improved vaccines against microbes for which vaccines currently exist but may not be useful for the civilian population;
- Develop new vaccines for microbes against which no vaccines currently exist;
- Establish needed research resources and make them available to the scientific community.

NIH's extramural research grants provide a mechanism to attract not only the academic community but also private firms. The money goes out in several forms: general grants, cooperative agreements, and contract resources. She highlighted the importance of SBIR grants to the small business community, which "really rallied after 9/11." Within about a month, she said, NIAID had put out a solicitation to the small business community, detailing exactly what was needed. This drew about 300 responses within a month. "It was a phenomenal expression of interest and capability and good application, with extremely thoughtful approaches."

She described a recent effort to develop new models. One was a challenge grant model, in which the government grant must be matched 50–50 by the grantee, to use in developing products. This had been refined with the realization that a balance of 80–20, or even 90–10 was more realistic for most companies, while maintaining the principle of industry participation. The institute recognized that the way industry does business is different from the way academia does business, and most of the institute's mechanisms had previously been based on academic models.

The institute also developed a series of contract infrastructures. One, a contract for vaccine products, was unusual for the NIH. It meant that in some ways, the institute was operating like a medium-sized manufacturing company, offering to do some screening, animal model evaluation, preclinical testing, and clinical trials, and to help the company submit an application for an investigational new drug (IND). The flexibility of this model contributed to its success, because each company was likely to have different needs. Some companies might have the money to achieve all these steps; most would not. Some companies would encounter stumbling blocks; the institute could look for ways to help the firm move past them.

She said that NIAID did recognize that this kind of partnership was an important vehicle for business, especially in biodefense. Companies may not have in their own labs such specialized resources as assays for anthrax or nonhuman primates to do an aerosolized challenge model.

She also described other forms of partnering—with non-governmental organizations (NGOs), for example. She said that many people and organizations had tried to fill some of the post-development gaps, such as guaranteed purchases. Various collaborators had also provided surveillance data or needs assessments for particular products.

She focused then on vaccine production as an "example of the kind of things we can do." Within each program in the biodefense research pathway, NIAID supported 30 to 40 to which it was encouraging people to apply. Each initiative was targeted toward a certain area, with its own specific approach and needs.

Models for Vaccines

Three specific models were evolving around biodefense vaccines. In the first, DHHS awarded a $428 million contract to produce a smallpox vaccine. This was a guaranteed-purchase model, which was a usual model for the government, and it inspired differences of opinion on its viability.

A second model was a "medium-sized pharma model" designed to build infrastructure. Since NIAID was created in 1962, it has been building and working with Vaccine and Treatment Evaluation Units for developing vaccines. This kind of infrastructure allowed NIAID to respond quickly to biodefense needs. Because the units were under contract to NIAID, the institute could arrange the priorities.

As the institute waited for its post-9/11 contracts to be implemented, it performed an inventory of smallpox defenses. It found available only 15 million doses of smallpox vaccine for a U.S. population of some 280 million. Researchers then asked themselves whether they could dilute the dosage, and found that the answer was yes: They could dilute it at least 1:5, and perhaps 1:10 if necessary. That raised the number of doses to 75 or 150 million. They also analyzed what could be expected in the way of clinical reactions if the entire population had to be vaccinated, which was very important for policy making. They found that in a normal population that is vaccinated with smallpox vaccine, there are many adverse reactions, but most are not serious or life-threatening.

In addition, NIAID was able to find additional stocks of vaccine that had been made by Aventis Pasteur three or four decades ago. They obtained this vac-

1900-
1904: Average of 48,164 smallpox cases annually in United States

1949: Last reported case of smallpox in United States

1967: WHO Smallpox Eradication Unit established

1971: Last smallpox case in Americas (Brazil); routine childhood smallpox vaccination ends in United States

1977: Last smallpox case in endemic area (Somalia)

1978: Two additional smallpox cases in laboratory accident in Birmingham, England

1979: Global Commission for Certification of Smallpox Eradication certifies global eradication of smallpox

1980: 33rd World Health Assembly of WHO officially declares global eradication of smallpox

FIGURE 17 Smallpox timelines.

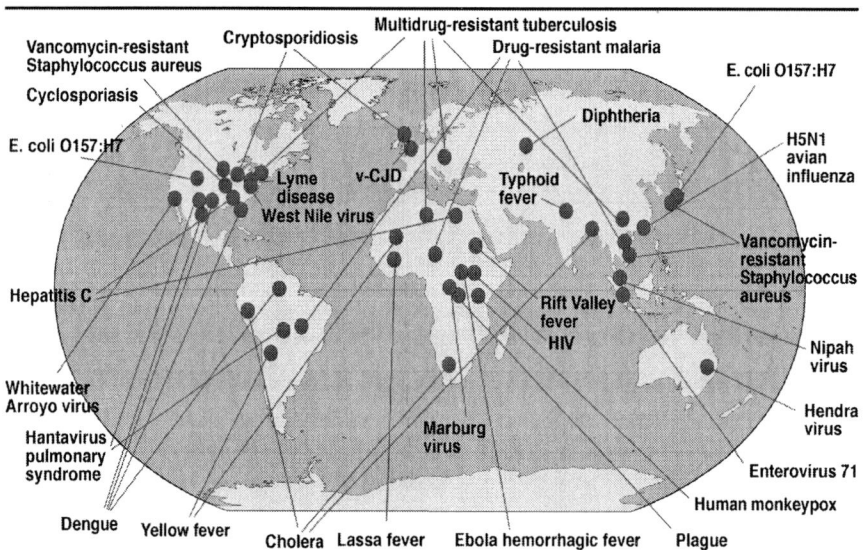

FIGURE 18 Examples of emerging and re-emerging diseases.

cine and found that it was still potent and could also be diluted. So there was now at least enough known smallpox vaccine for the entire U.S. population.

The third model, designed to develop a vaccine against Ebola virus, had been developed over the past four years at an intramural facility called the Vaccine Research Center in Bethesda, Maryland. For Ebola, NIH formed a Cooperative Research and Development Agreement Opportunities (CRADA) partnership with a small company to produce and market the vaccine, and determined what would be the most useful assistance NIH could provide in order to make the partnership work.

As her last point, Dr. Heilman announced a piece of breaking news. Just 4 hours previously, she said, NIAID had announced a new contract for the development of an anthrax vaccine. This would be different approach, using a recombinant protective antigen, in which researchers use a fragment of the anthrax agent instead of the whole organism as the antigen. From preliminary studies performed jointly with DoD for the past 5 years, the institute had concluded that the product had "a good safety profile and seems to have a nice antibody profile." Now that the decision had been made to proceed with the technique, she said, this particular program would be fast-tracked.

She concluded by reaffirming NIAID's support for public-private partnerships. "Our job is not to market vaccines," she said. "It doesn't make sense for us to develop vaccines if we have no partner at the other end to bring them to market. So for us, a very important component in deciding how heavily to invest in vaccines is the commitment of the private partner."

PARTNERING FOR COUNTER MEASURES: THE PRIVATE RESEARCH PERSPECTIVE

Gail Cassell
Lilly Research Laboratories, Eli Lilly & Co.

Dr. Cassell said she would talk about three different topics: (1) the current status of countermeasures against biothreat agents—focusing not on vaccines, which Dr. Heilman addressed, but on antibiotics and antivirals; (2) the partnerships that had already been established and seemed to be working; and (3) issues that may need to be resolved to encourage more partnerships for future development of countermeasures.

With the diversity of biological weapons and the ever-increasing possibilities to create new weapons through genetic engineering, she said, there was no simple way to develop countermeasures to biothreat agents. A sobering consideration was that it might take only 3 to 6 months to develop a new biological weapon, but at least 8 to 10 years to develop a new antibiotic, antiviral, or vaccine.

She also pointed out that because of the diversity of biological weapons, including a number of different viruses and bacterial agents, many infectious agents require broad-spectrum therapies.[14] The same holds true with respect to development of anti-viral treatments. In the past, research on anti-virals had focused mostly on agents for specific viruses in a one-to-one relationship. Researchers would like to be able to use new technologies to develop broad-spectrum antivirals for use in emergencies. To date, this had not been possible, she said. In our armamentarium today, we had 13 different viruses on the "select" list, but only a single anti-viral was being tested, cidovifir,[15] a derivative of which was indicated for treatment of vaccinia and other pox viruses. Even this drug could only be administered intravenously, limiting its usefulness in time of emergency. In addition, it is nephrotoxic, as antivirals often are, which may limit its safe use in children.

In the face of the large number of major viral diseases, health officials had at their disposal only a small number of anti-virals to naturally occurring viruses. The only antivirals that had been approved and marketed for use were those used against HIV, hepatitis B, and herpes.

For antibiotics, the situation was somewhat better, she said, but still worrisome. Over the last 30 to 40 years, only two new classes of antibiotics have been developed and introduced. The more recent was approved several years ago, but unfortunately, resistance to it developed even before FDA approval and launch. Before that—about a decade ago—companies were excited when genetic sequencing was completed for a number of important bacterial pathogens; this heralded two decades of drug development now regarded as the "golden age of antibiotic discovery." Again, however, results have been disappointing. Even with millions of dollars invested in trying to develop new classes of antibiotics, the products that had reached phase one through three development, and even late preclinical stages, still fell short of completely new classes of antibiotics or new broad-spectrum antibiotics.

At first glance, a tally of the antibiotics that had reached phase one through three development seemed encouraging, she said, with a total of some 20 new drugs. But a closer examination would reveal that these were not truly new classes of antibiotics, but slight modifications of tetracycline and other macrolides and a larger number of quinolones. Quinolones can be effective drugs, she said, but bacteria develop resistance to them quickly. In China, for example, some 50 per-

---

[14]A narrow-spectrum antibiotic might be effective against one or several bacteria or other agents. A broad-spectrum antibiotic might be effective against a group of organisms, such as gram-positive bacteria, or a vast array of different organisms. Overuse and misuse of broad-spectrum antibiotics has been associated with the emergence of antibiotic-resistant strains of bacteria.

[15]Cidofivir itself is licensed to treat infections of the retina caused by cytomegalovirus in patients with HIV/AIDS. A derivative of cidofivir has stronger anti-pox activity.

cent of the strains of *E. coli*, the common intestinal bacterium, are already resistant to commercial quinolones. A new class of broad-spectrum antibiotics is needed, because no one knows what specific agent might be used in bioterrorism. The best strategy, she said, would be to develop a drug with broad-spectrum activity to hold in reserve for immediate response; that drug could then be replaced by drugs that are more selective when the agent was identified.

To summarize her discussion of antivirals, she emphasized there exist virtually no broad-spectrum anti-virals in phases one through three in the development pipeline. She did say that a few immune modulators hold promise, but no broad-spectrum antivirals.

The Puzzling Failure of High-Throughput Screens

She described a puzzling technical challenge that stands in the way of both broad-spectrum antibiotics and antivirals. There are many companies performing high-throughput screens that contain over a million different chemical entities, and yet they have not found new effective agents. "What we need is a scientific explanation for that," she said. "It's not the targets; these are numerous and well validated. The problem is that we are not finding new chemical entities that can actually inhibit the growth of the organism."

There are also financial hurdles, said Dr. Cassell. The failure rates for drug discovery, in areas from organ identification to launch of a new drug, average about 90 percent. In the area of antibiotic drug discovery, the failure rates had been even higher. This reality had dampened enthusiasm among firms for investing in antibiotic drug discovery. In fact, she said, a recent competitive analysis indicated that most large pharmaceutical companies and many smaller companies had actually reduced their activity in this area. Because of the daunting technical and financial challenges, they had emphasized therapeutic areas with unmet medical needs and larger market opportunities.

Another reason companies have shifted away from the complex challenges of bacteria and viruses is the new pharmaceutical opportunities opened by completion of the human genome. Geneticists have shown that researchers have been focusing on only about 10 percent of the potential targets for drug discovery in fields such as cancer, endocrinology, and neuroscience. Now there were hundreds of new potential drug targets in the form of genetic segments. Drugs with potential activity against specific gene targets are less complex and far easier to develop than those needed to inhibit the growth of bacteria or other natural products.

In short, she said, what we have now in our antibiotic and antiviral armamentarium is quite limited compared to the ideal. The picture is also clouded by concerns about increasing antibiotic resistance. Despite a great deal of public and Congressional attention, it is clear that no public health response to bioterrorism is likely to prove effective without (1) addressing the overall problem of antimicrobial resistance and (2) making better progress against both technical and financial challenges of drug discovery, both for bio-agents released intentionally and those occurring naturally.

The need for partnerships is no greater in any area of bioterrorism, she said, than for developing new countermeasures for biological threats, particularly in the areas of antibiotics and antivirals. She regarded these tasks as more challenging scientifically than development of vaccines, where "in some cases you can get away with rather crude vaccines that provide excellent protection. Our easy antibiotics have already been discovered, as well as the antivirals."

In the face of these challenges, she said, partnerships are crucial. The very best and brightest scientists were spread throughout NIH, the universities, and the pharmaceutical and biotechnology industry. No single institution or sector could address the challenges alone. In addition, partnerships were needed in order to share the high financial risks, particular in the area of antibiotic drug discovery, and to pool dispersed knowledge about health risks.

Testing the Utility of Existing Drugs

She then turned to some partnerships that had been established since October 2001, some initiated by Lilly, others through various pharmaceutical companies. One was a working group with representation from the NIH, DoD, CDC, and FDA, charged with assessing the utility of existing antibiotics and antivirals. The importance of such an exercise, she said, was illustrated by the usefulness of older antibiotics when the anthrax attacks took place. Older antibiotics are likely to have efficacy against the agents discussed by Dr. Heilman, but proof is needed, both in vitro and in vivo, and other questions must be addressed. For example, what kinds of standards are needed to demonstrate in vitro activity against intracellular organisms? Can this be shown in routine laboratory facilities, or does one have to do susceptibility tests in the presence of cells? When does efficacy in vitro demonstrate efficacy in vivo? The last question is complicated because it is not ethical to perform clinical trials or even experimental studies on humans. There is little experience with naturally occurring human cases or the use of antibody therapy, particularly for organisms on the select list.

The FDA uses a rule stating that efficacy based on animal data must include at least two different species before it can be accepted as a standard. A problem for some bacterial agents and virals, she said, is that there are no accepted animal models. Researchers had begun the task of determining what standards to use and what body sites were the correct sites to use experimentally. In addition, a vaccine working group was being established, composed of companies that produced vaccines, to decide which viruses to focus on in developing new vaccines and to assist NIH-funded investigators. Industry investigators had also agreed to share information and data in other ways that might point to promising directions or avoid known dead-ends.

She also described an information technology initiative that would take advantage of expertise in the pharmaceutical industry and in IT companies such as IBM. This group would advise the WHO, CDC, NIH, DoD, and other entities in establishing a global surveillance network, with special coverage of certain countries. The DHHS had committed $10 million to this effort, and others were committing resources and expertise. Such an international effort was important, she said, because modern travel and trade systems can transport microbes to and from every region.[16] She noted that recent international discussions on weapons verification had emphasized surveillance as one of the most important components of any system to control biological weapons. One participant in these discussions, she said, was a training program for international fellows established by Lilly at the CDC, which had recently been expanded to included fellows working on biothreat issues.

Another partnership established between HHS and CDC, she added, distributed educational materials about the highest select threats. This included hand-delivery of protection materials by pharmaceutical sales forces to practicing physicians. HHS regarded this as important because it placed practical information at the point of delivery.

Important Issues to Address

Some important working groups had been established, she said, while others needed to be formed and nurtured. These must address some complex issues that had not been resolved. For example, the nation still had significant vaccine shortages even for normally occurring childhood diseases. Many other issues remained to be addressed around the issues of antibiotic drug discovery, liability, and indemnification. The question of liability, in particular, is extraordinarily complex in relation to human health and biothreat agents, both because of the animal model rule and the inability to gather sufficient data to show safety and efficacy in humans. If anti-trust issues that preclude company consortia could be resolved, the resulting partnerships could provide working relationships that allow risks and expertise to be shared. Dr. Cassell said she served on the Global Alliance for Tuberculosis, and even that forum is hindered by antitrust issues from determining the manufacturing capacity of existing antibiotics. One of the most thorny issues to be resolved was that products related to bio-agents have a limited and undefined market. Collaboration will be required to develop the ability to project the real costs of producing and safeguarding products that may be needed in the future.

---

[16]Her cautionary words preceded the 2003 outbreak of severe acute respiratory syndrome (SARS) in China, which spread rapidly to Hong Kong and then to Toronto by way of infected airline passengers.

In addressing all these challenges, she said, government incentives and contracts needed to be realistic about the costs required. "The important increase of $1.5 billion to NIAID," she said, "is really only a drop in the bucket to what's needed when you consider that it costs almost $800 million to develop a single product." She pleaded also for realistic estimates of the time and vast amount of research needed to develop countermeasures. Any legislation, she said, should also encourage the broadest participation to ensure the best products and the lowest cost. "Excluding or disadvantaging some sectors of the industry," she said, "would work against this goal."

She ended by concluding that the best deterrent against the use of a biological weapon of mass destruction may be a constant stream of new, innovative antibiotics, antivirals, and vaccines. "Knowledge of such commitment and successful development would surely help dissuade our enemies in such an arena," she said. "But successful innovation and development will require successful and effective partnerships."

DISCUSSANT

Kathy Behrens
RS Investment Management

Dr. Behrens reminded the participants that there is a "silver lining for this set of storm clouds": All the work we do on bioterrorism will also contribute to understandings and solutions to natural causes of disease and infection.

She began by saying she would reinforce some of the reasons in favor of government-industry partnerships. For the area of bioterrorism, she said, "the good news is that we have a strong and favorable history of partnerships at many levels"—government-academia, academia-industry, industry-government—where many organizations and enterprises were already working together. She said that she hoped the urgency of the task ahead would help to reduce some barriers that in the past had slowed the development of some drug products. She said that by the evidence presented by the panel so far, "it's very clear that a lot of agencies, private organizations, and academic enterprises have put pencil to paper in trying to solve these problems, in spite of the fact that you can't tell from day to day."

She expressed optimism about partnerships on the basis of what the STEP organization had done through its series of conferences. She said that the group had learned that one key to partnership success was to have well-defined objectives and goals. In the bioterrorism area, she said, whether for protection, vaccine manufacture, or passive immunization, "it's clear that we have a target set of organisms we need to be working on and some very specific results we're looking for to provide protection for individuals."

The first step, she said, had already been identified historically for government-industry partnerships, and was already in place. It began with personal interaction and good engagement between the individuals who represent different entities. She reminded her listeners that all organizations established for the purpose of providing advice should feel reasonably sure that they are there because their viewpoints are respected, and because people want their feedback and advice.

Key Areas: Liability and Regulation

She suggested that the panels look further at some of the areas discussed by Dr. Cassell, especially in determining the kinds of projects government-industry partnerships are equipped to do, discovering whether the dialogs that will be necessary in initiating partnerships have yet begun. Key items, she suggested, were the broad areas of liability and regulation, which had been major contributors to the cost and time of development for many therapeutic agents. She stressed the importance of good communication with the private sector, including assurances that these issues would be addressed thoroughly during the design of any partnership.

Dr. Cassell then joined the discussion, affirming that those issues had indeed been raised and discussed over a number of years—usually in regard to the development of vaccines that would be used in developing countries. The issue of liability had received serious attention in regard to infectious diseases. About 15 years ago, it was noticed that as drugs became more effective at eliminating infectious diseases, the reactigenicity associated with existing vaccines became a contentious issue. These examples, she said, could serve as precedents to build on during discussion of the same issues in the bio-defense arena.

She elaborated on the issue of liability. As vaccine safety became more of an issue, it prompted the development of a vaccine compensation program. That program had allowed for vaccines to be taxed, and for that tax revenue to flow into a compensation program. As adverse reactions were identified, by almost anyone, sufferers were able to go to the compensation program for relief instead of to the legal system. The value of this system was that during the mid-1980s, when law suits became prevalent, fewer vaccine companies were lost than had been feared. Therefore, the model of compensation was being studied as a one that might be used in developing and using bio-defense agents.

In addition, she said, the Department of Defense had had experience with liability issues in developing their vaccines. Most of the test vaccines were used under IND (investigational new drug) status, and the authorization for DoD already contained liability capabilities. She said that she believed this kind of authority already existed within the DHHS, although it may not have been exercised yet.

In the area of regulatory issues, Dr. Cassell pointed to "another silver lining" in government-industry relationships: "We not only knew each other before, but have trusted each other and worked together." This was true both for DHHS and its regulatory counterpart, the FDA. She described a productive precedent in a model used for the accelerated development of acellular pertussis vaccine. Under that model the FDA, NIH, CDC, the DHHS, and the drug companies came together periodically "and resolved issues then and there." She said that this model was then being used for development of improved smallpox vaccine by two private firms[17] and that a similar model was being implemented to develop rPA vaccines against anthrax.[18] The model benefits from the use of simple conventions, such as entering data in a format used by the FDA; this saves the time that might have been spent re-entering data into a different format before it could be evaluated.

She said that the animal model rule would probably be used for the first time on bio-defense drugs, and the FDA had proposed a different paradigm for approaching it. The customary FDA procedure had been to ask industry to first lay out a plan that the agency would then respond to. In the new paradigm, the agency invited companies to sit down together to define what needed be resolved at the outset. The agency would suggest the guidelines, but invite the companies to discuss any problems they might have with those guidelines. "Those kinds of activities are ongoing now," said Dr. Cassell.

The Difficulty of Using a Procurement Model

Dr. Behrens asked whether a procurement model would be the best vehicle for ensuring adequate return for the industry partners. Dr. Cassell agreed that the procurement model already used for vaccines is an important one to study. She said that a problem in using the model with antibiotics and other therapeutics was that it may not be possible to project the need as accurately as it is for a vaccine, where a fixed dose gives protection. She said that the model would have to be studied further with respect to many variables, such as what problems does the model currently present; what will be the source of funding for procurement; how stable will the drug be; and how variable might the need be over the next decade.

---

[17]In January 2003, HHS awarded two contracts totaling up to $20 million in first-year funding to develop safer smallpox vaccines. The three-year contracts were awarded to Bavarian Nordic A/S of Copenhagen, Denmark, and Acambis Inc. of Cambridge, Mass. The National Institute of Allergy and Infectious Diseases (NIAID) administers the contracts.

[18]rPA is the abbreviation for anthrax recombinant protective antigen vaccines. In this model, NIAID announced a partnership agreement with a private vaccine production and support contractor, Science Applications International Corporation (SAIC). SAIC was designated to solicit solutions and serve as the main contact point for information about potential (rPA) vaccines from all sources.

"In the case of therapeutics," she said, "there is a finite shelf life, so we have to know if it will need to be replenished."

Dr. Heilman said that these issues had indeed been brought to the attention of the department, and she acknowledged that any model would have to be complex if it is to address the many variables already anticipated. "The good thing," she said, "is that there's a dialog and we're trying to resolve the challenges. The hard thing is that there's a lot of complexity in finding an answer that can indeed be valuable for everybody."

Dr. Behrens continued with the topic of finding the best match with industry in planning for research and possibly development activities. The good news, she said, is that there are a small number of players, and the relationships between them and government had always been good. But, she asked, was there a mechanism that will allow us to match up concepts with both small and large entities that are best able to conduct the work, and an efficient way to bring the parties together?

Dr. Heilman said that the agency's Web site had been very helpful. She assured the panel that NIAID had a 40-year history of vaccine development and it was familiar with every competent participant. She also applauded Dr. Cassell for working tirelessly on behalf of the industry as a whole to optimize the interaction between government and industry. She cited the example of Dr. Cassell's extensive effort to understand the full scope of industry's research on adjuvants,[19] so as not to repeat work already done or to follow known dead ends during new programs. She noted that the agency benefited from good relations with PhRMA and similar organizations, which allowed government agencies to communicate their priorities directly to responsive industry representatives.[20]

Dr. Behrens added that efficient communication is essential when there are so many pending actions in different places, all on "parallel tracks. The right people have to be in the room to bring these tracks together."

Praise for the New "HSARPA"

Dr. Kerr said that the new Department of Homeland Security planned an entity called HSARPA, the Homeland Security Advanced Research Projects Agency, under the Undersecretary for S&T. HSARPA, he said, would be "the systems equivalent of DARPA, but with many of the procurement issues and problems put aside." HSARPA would be the "major facilitator to couple the research and development testing and evaluation enterprise with the actual entities, whether they be in the private sector or in academia, and the actual end-users."

---

[19] An adjuvant is an immunologic agent that increases an antibody response.

[20] PhRMA (pronounced pharma) is the Pharmaceutical Research and Manufacturers of America, which represents the country's leading research-based pharmaceutical and biotechnology companies.

This facilitation will occur in the entity that reports directly to the undersecretary.[21]

A questioner volunteered that the idea of HSARPA was an excellent one, but said that the published budget figure of $200 million seemed far too low, given the agency's responsibility for both animal and human clinical research. Dr. Kerr replied that the initial budget should be thought of as "an administrative setting-up period in which the actual roadwork and technological and administration will be set in place."

Dr. Behrens praised the job of Dr. Cassell in identifying the historical issues on the pharma side, especially in the case of anti-infective and similar agents. She then asked Dr. Heilman whether industry should be doing anything now that was not being done. Dr. Heilman answered that she could not think of an example, but that "people seem to be coming together on this issue more so than around any area I've seen."

A Need for More Manufacturing Capacity

A questioner asked about procurement, and how long it might take to create a new manufacturing facility. Dr. Cassell answered that the point was important. For even the existing antibiotics, she said, there is little or no excess manufacturing capacity. If the nation had to gear up to meet a surge in demand, it would probably require months, not days. This is true for biologics, because the manufacturing is so complex and the quality control standards are high. She said that at a meeting at the National Academies in December 2002, the FDA admitted the possibility that it might have to build a new facility for manufacturing biologics. She had concluded that constructing a dedicated facility, which would have to be maintained at high standards even it was not being used, was not cost efficient. At the same time, she said it was difficult to say how a demand surge for anti-infectives could be met without sacrificing some needed product for which there was continuing demand. "Do you take away antibiotics that are needed to treat sepsis over here because we have a biothreat over there?" she asked. "Those are not easily answered questions."

Suspending Patents Could Have a "Chilling Effect"

Steve Merrill of the National Research Council returned to the issue of liability. He recalled that HHS Secretary Thompson had raised the possibility of abrogating the Bayer patent on the antibiotic Cipro if that was necessary to obtain an

---

[21]The model for HSARPA, the Defense Advanced Research Projects Agency of the DoD, is respected among government entities for its flexibility and innovative thinking. Aspects of the original Internet, for example, grew out of research funded by DARPA.

adequate supply of the antibiotic for an emergency.[22] At the time, he said, some warned that this comment could have a "chilling effect" on companies' willingness to develop new antibiotics, vaccines, and anti-virals. He asked if such an effect had occurred, and whether the government had developed a position on protecting such intellectual property rights.

Dr. Heilman said that in order to have success in this area, the government would have to set a policy environment that not only would nurture public-private partnerships, but also would provide strong intellectual property rights. Without them, she said, companies would not support the innovation that is required. This was a lesson that had been learned through the experience of the vaccine industry—that innovation is essential, "especially if we take into account the National Security Council's warning that infectious diseases in general are the most serious threat for humanity over the next two decades." The finding of this report, released 3 years ago, had become even more urgent, she said, because it was written before the anthrax attacks.

Dr. Behrens said that in small and mid-sized companies the topic of liability is raised more often than any other. "The secretary's comment really struck fear in the hearts of organizations that finance the business," she said. "It was resolved reasonably amicably, but had the potential to cause serious harm."

Dr. Heilman agreed that questions of rights and liability pertained not only to single-use products developed for select agents, but also to multiple-use products like ciprofloxacin.

A questioner asked Dr. Heilman whether sufficient numbers of trained people, both in government and industry, were available to work on bioterrorism research, given the fifteen or so years required to train a first-class biotechnology researcher. He also asked whether the specific skills needed to combat terrorism were being taught at all, and whether there needed to be incentives for candidates.

The Lack of Trained Biomedical People

Dr. Heilman said that that was in fact her largest concern—not only are there insufficient numbers of people now trained in microbiological and immunological sciences, but there are few incentives to attract them away from other important research. She recalled a recent discussion about workforce issues related to bio-defense. She said that available data about immunologists, especially those well trained in the clinical microbiology needed in surveillance and research in public health, indicated the existence of "maybe three people" in the United States with expertise in plague and anthrax.

She also said that the nation lacked sufficient numbers of in vivo biologists—people trained in whole-body physiology. The need for DVM/PhDs and

---

[22]Ciprofloxacin is a broad-spectrum quinolone approved for use against inhaled anthrax.

veterinarians is a major issue, she said, not only to help establish infectivity models, but also to address animal diseases and agroterrorism. "The manpower issue is tremendous," she concluded.

The government is making efforts to address this issue. Dr. Heilman said that DHHS does have a targeted initiative for training in the area of bio-defense. A second effort was to identify regional centers of excellence that would partner with public health service systems within the region and with the CDC. Those partnerships could be regarded as additional surge capacity, she said, should the need occur. "It's better to partner now than during an event," she said, adding that the partnerships would also form units to train people on site in various skill and techniques.

Marc Stanley, director of the NIST ATP program, referred to the long history of the ATP in sponsoring government partnerships with both industry and academia, and suggested the use of this vehicle for bioterrorism as well. "If the effort of the government is to get advanced technology commercialized quickly," he said, "it seems to me having a viable program tested over eleven years would be a unique way to utilize those capabilities without having to reinvent the wheel." He also suggested that the effort make use of existing relationships in the DNA diagnostic field between NIH, NIST, and Sandia Laboratories.

Dr. Heilman responded that a partnership for DNA sequencing had already been developed along the lines the questioner suggested, with the goals of sharing sequencing information, identifying priorities, and finding the best people.

With regard to the question about NIST, she said that NIAID was working with a DoD organization called the Chemical, Biological, and Radiological Technology Alliance (CBRTA). "In some ways they have been able to solve the IP issue," she said, "and have allowed interaction along the whole spectrum of bio-defense. It's a fascinating concept to see whether we can interact with them as they're doing their mission for the DoD."

A questioner referred to the expense of developing drugs by current processes, and asked whether new techniques such as robotics would reduce the cost and time of drug development, and what public-private mechanisms might advance development of these new robots. Dr. Cassell said that the "short answer is that it hasn't yet reduced the price, and the reason is that the new technologies being used are more expensive. Unfortunately the science is expensive, and biologics, because they are more complex, are more costly." She said it is also appropriate to focus on the value of the outcomes—how many lives will be saved and how much health care costs decline by applying new discoveries. She did say that the failure rate in the drug discovery process is likely to decline because of new data from microbial genome sequences, the human genome, and pharmacotoxicology. "We'll be able to better predict potential toxicities before we actually try a new substance in animals and humans. But we're not quite there yet."

Panel IV ————————————————

Partnering for National Security

INTRODUCTION

William B. Bonvillian
Office of Senator Lieberman

Mr. Bonvillian, who had been involved in shaping the homeland security legislation, conceded that it had been a "struggle," at least partly because it was located "in an unclear legislative area." He suggested that the panel could help answer one of the most difficult design questions—how to create the government-industry interaction that is necessary to stimulate the innovations that will be needed by the Department of Homeland Security.

One reason to hope this might be possible, he said, was that Congress had succeeded in crafting an apparatus within the legislation that supports scientific and technological innovation. On this, he said, Congress had showed a "remarkable and bipartisan consensus" that was "heartening."

In countering bioterrorism, he said, the largest research portion of the R&D challenge is on "the biothreat side." In the physical sciences, the issue is more developmental: the need for better sensors, detection, knowledge management, data mining, information technology, and physical protection.

He listed some of the other major challenges:

• While the R&D challenge is major, the deployment challenge is "huge." Some 85 percent of the nation's critical infrastructure is in private hands, he said; this meant that "technology transition is going to be crucial for this new department."

- The new department had a big information management problem. The department is large, and it needs to exchange information with the other agencies it interacts with, and those agencies represent a large part of the federal government.
- Most of the R&D on homeland security would continue to be performed outside the new department. The contributions of DoD, NASA, DoE, and NIH would dwarf what occurred in the new department. The challenge was "get all these players operating on a common R&D roadmap."

Key Elements in the New Department

He reviewed some of the key elements in the new department. First, the S&T activities were headed by an undersecretary. It was necessary to have an individual with sufficient rank and title, he said, to gain the attention of peers in this complex interagency process. Second, the new department was designed to include a DARPA-like entity to emulate one of the roles DARPA[23] had played: to use its funding to leverage contributions from service R&D and from R&D centers throughout the defense research establishment. He credited DARPA with "playing a coordinating role and encouraging concentration and focus within DoD." That, he said, would be the objective for this one significant piece of the Department of Homeland Security. "It will have DARPA's leanness, its flexibility of personnel, access to talent, and freedom from some Civil Service limitations that are not appropriate for a scientific organization—in other words, the procurement flexibility that DARPA's used so effectively over the years." He said that HSARPA would also have an acceleration fund, with a "significant" level of funding. The goal, he said, was to enable it to leverage participation and cooperation across a whole series of agencies, in addition to what the department is undertaking, and also to involve the private sector.

A third element of the DHS research plan was the use of the FFRDC[24] concept. The undersecretary of S&T would be able both to establish an FFRDC and to have ready access to existing FFRDCs. One of the central recommendations by the National Academies in their homeland security conclusions was the recommendation that an FFRDC should play a key role, particularly in the areas of threat and risk assessment and risk management.

---

[23] See also Dr. Kerr's comments about HSARPA.

[24] A Federally Funded Research and Development Center (FFRDC) is a unique organization that assists the United States government with scientific research and analysis, systems development, and systems acquisition. FFRDCs bring together the resources of government, industry, and academia to solve complex technical problems that cannot be solved by any group alone.

Other elements, he said, remained to be worked out. One was a mechanism to coordinate among the various agencies that would be involved. The Senate bill proposed a coordination council, with statutory framework, tasked to develop an interagency technology roadmap for homeland security. It would also have private sector participation, advice, and support.

Another element to be added was a mechanism that served as a clearinghouse. One of its jobs would be to evaluate private-sector technology solutions. After the 9/11 attack, he said, DoD was inundated with technology suggestions that it had still not been able to sort and respond to. A clearinghouse was needed to manage, identify, and evaluate national technological opportunities that might be relevant to the new department's and the nation's needs.

A third job for the new entity would be to play a significant role in encouraging and sponsoring technology transition. In addition to the problem of coherence in technology development outside the department, he said, there is "a huge problem inside." The DHS was enlisting the collaboration of numerous entities from different agencies in the government that did not have a history of close cooperation. DHS will need to have them to cooperate, and to have interoperable systems in order to function efficiently.

A Lack of Good Models

The implication was that these entities would have to pair up more closely than they had in the past. Mr. Bonvillian described several models that had already been tried, with less than perfect outcomes. In one, DoD had managed the different jurisdictions responsible for ports, aviation, and border security by joining acquisition and technology beneath the same undersecretary and paralleling that structure with assistant secretaries in the services. A criticism of that model has been that acquisition became the main focus, with technology acting as the "bill payer," in effect, for acquisition activities. For DHS, he said, a different approach would be tried, making the S&T function more independent. But how do you get technological coherence if you separate S&T? he asked. "By having the undersecretary for S&T serve as a technology officer for the department, in a business sense," he said, "and ensuring that the undersecretary has the sign-off on the testing and evaluation process." This would create a clear point well before procurement where technology, interoperability, and connections are ensured.

He said there was also an "understood" strategy for private-sector involvement at an early stage. The challenge was that pharmaceutical and biotech companies, which make vaccines and other drugs, need a market incentive to justify R&D investments. The possibility of a public health disaster was not an incentive that would draw a biotech firm into the biothreat business. The government needs to help firms develop business plans and find new incentives, such as intellectual property incentives and procurement incentives. "I would argue," he said, "that there are other areas where that same kind of creative thinking is going to have to go on in the private sector."

He noted that this was the first significant new science and technology agency the government had created in 45 years. The government had generally had a policy of decentralizing science, which had brought many advantages. There is no "science czar" or science ministry; instead, there are many actors pursuing different missions. There are advantages to this diversity, including the research power of multiple points of view and approaches, the opportunity for research entrepreneurship, and the good chance of avoiding the mistakes that often occur when a centralized science bureaucracy dominates decision making.

The downside becomes apparent when the government needs to build cross-agency and cross-discipline collaboration, as it does in the case of homeland security. "We're in an era where advances come out of cross-disciplinary work," he said. "How do you create those connections? We haven't been able to do that very well in our science portfolio." The new DHS, which had a bipartisan consensus, would try to deal with that problem by using the acceleration fund as an incentive for interagency cooperation. It also planned to use the interagency council and a technology roadmap from that council.

Better Management of Technology Transition

Mr. Bonvillian also addressed the underlying issue of technology transition, because of the need to have the full cooperation of the private sector. He described a clearinghouse mechanism to enhance private-sector participation, with the goal of a non-bureaucratic entity that is lean, entrepreneurial, flexible in its procurement practices, and has access to quality talent. Obviously, he said, one could not "write culture into a statute," so much would depend on the original leadership of the department. A helpful beginning would be to enable private-sector leadership rather than government leadership in critical areas.

He concluded by pointing to the need to "tackle the whole information management challenge within this department and connect it to other entities—for example, in the area of intelligence." He predicted that the new department would be one of the largest collectors of information in the government—"staggering amounts of information coming in about people and goods and cargoes and items entering and departing this country." That information, he said, would be worthless unless it connects with the knowledge of threats or can be analyzed to develop information about threats. This would require efforts to build strong information management, and to integrate the department's R&D, including effective connections and opportunities with the private sector.

"That's a background of what we're aiming for in this new department," he said, "and some of the challenges that Congress has been addressing."

OVERCOMING INFORMATION OVERLOAD

Anne K. Altman
IBM Corporation

Dr. Altman, who manages IBM's relationships with the federal government, said that the "overwhelming amount of information" that many organizations, especially government agencies, must deal with today constitutes a daunting challenge. She said she would talk about some of the insights she and her colleagues at IBM had gained from working with other large organizations with information challenges.

She divided those insights into three areas. The first was the need to develop an integrated information architecture—one that actually serves to achieve its objectives. The second was to create partnerships to collect and manage information—not just government sharing with government, and government with academia, but government with business and government with citizens. The third was to implement technologies and public policies that ultimately enhance the government's ability to partner and achieve its missions.

She opened with a brief discussion of the new Transportation Security Administration to illustrate "just how overwhelming the information challenge can be." She described the multiple data exchanges between the government and the private sector generated by each passenger who boards a commercial airplane in the United States. She saw this process as a "powerful statement of the direction and vision of coordinated information." What the process illustrated, she said, was the future of passenger transport and security that is TSA's mission. The TSA describes its mission as ensuring the freedom of movement of passengers and cargo.

Every day, Dr. Altman said, some 1.6 million passengers board commercial airplanes in the United States, bringing three million pieces of luggage to 429 airports.[25] She began with the private-sector data. For each passenger, data are collected by the travel agent, the airline, a rental car agency, perhaps the airport authority, and a credit card company. Information collected by the government might include passenger lists (which in the future would be pre-screened and compared with law-enforcement databases); information from the FBI, the INS, and perhaps the CIA; perhaps a tip from a citizen that is "used and massaged" in evaluating a passenger profile. In addition, carry-on bags are scanned; checked bags are examined for explosives; and video cameras record passengers' movements in strategic locations.

She noted that airport security is just one component of the Borders and Transportation component of the homeland defense mission. The additional five

---

[25]She said she had learned the previous day that the number of airports had been amended to 446 airports, which served as an illustration of the complexity she was describing.

components are Intelligence and Warning, Domestic Counterterrorism, Infrastructure Protection, Catastrophic Threats, and Emergency Preparedness and Response. And each component must have an information strategy that is integrated with the strategy of every other component.

In the case of port security, for example, she said that U.S. ports handle 16 million containers annually, or 44,000 containers a day, delivering nearly half of all imported goods. Across its borders, the country admits every year some 11 million trucks, 2 million rail cars, and 330 million legal visitors. For all of this flow of people and goods data must be collected and managed.

For the new DHS, the six primary missions are to be accomplished by 22 organizations described in the legislation (see Figure 19). Each organization must coordinate people, processes, and information structures and systems. The chart demonstrates the possibility of a common information strategy across the six homeland defense missions.

She said that it was likely that the same information could be used to achieve multiple objectives across this organization, so that an overarching architecture must exist to make critical information available to execute this mission. There were important principles to follow if the U.S. is to be the world leader in using

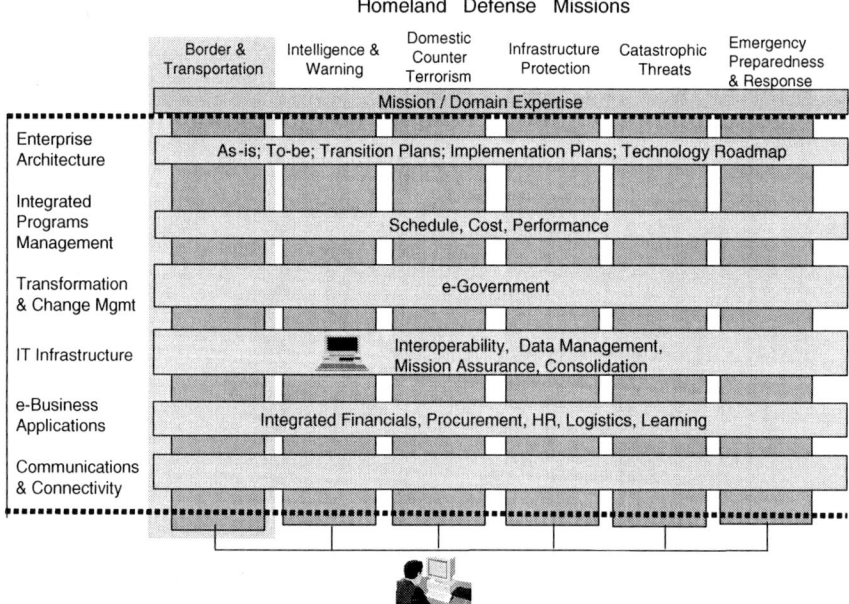

FIGURE 19 Partnership opportunity—Apply common information strategy across missions.

information to achieve Homeland Security. Federal agencies must not only work out on how they will share data among themselves, but also with local, state, and foreign governments, with the private sector, with academic institutions, and with citizens. She said that the challenge was much like that encountered during a merger in the private sector, "although I think this is perhaps the largest and most expensive merger ever attempted." Nevertheless, she suggested that policy leaders should draw on the experience of the private sector, which had successfully merged people, processes and information systems in ways that not only minimizes the cultural impact, but maximizes the overall effectiveness of the merger.

Two Keys to Successful Partnerships

There are two keys, she said, to successful partnerships. One is technology and one is public policy.

For technology, she discussed her experience in meeting with customers after September 11. At major federal agencies, she discussed their concerns about how to respond and be proactive in the future. Three questions came up: the first was how to get old and new data systems to communicate. The second was how to link together information that resides outside a particular organization. The third was how to link and connect data from a variety of disparate data sources.

She illustrated a tool called WebSphere, a type of software called middleware, that had been developed to address these concerns. This tool is designed to enable systems to communicate and conduct a wide range of information services and transactions. In the middle was a "discovery link," which creates a "federated data base," developed by IBM in the life sciences arena to help in drug discovery. This technology could enable agencies to connect disparate data sources—text, video, audio, and other types of data—as if they were part of one common database.

Lastly, she discussed "Web Fountain," an advanced information discovery system that allows users to locate information that resides outside their organization. "What's exciting," she said, "is that it's a blending of the known information you have with information perhaps out on the Internet; it allows for rich trend, pattern and ultimately relationship detection, a powerful tool for the needs of homeland security."

Expanding government's ability to share and access data does raise many questions about privacy and security, she said. Law enforcement sources cannot be compromised, and companies have to be assured of unfettered access to sensitive data. Technology offers many of the answers. In fact, she said, IBM has a technology that can be employed to protect the anonymity of sources of information, which may facilitate the ability to share sensitive data between organizations.

But data security and privacy require much more than technology, she suggested. They require thoughtful policy. She said that policy decisions by lawmakers and federal executives are the foundation for long-term success. She listed public policies that are key to successful technology partnerships:

- Open, standards-based infrastructure
- Government as early adapter
- Review restrictions to data sharing
- Security and privacy
- Cultural change and leadership
- Critical skill sets
- Indemnification for business
- Research and development
- Investment in information technology

The Argument for Open, Standards-Based Platforms

She highlighted two of these, beginning with open, standards-based infrastructure.[26] The development of an information architecture or framework is critical, she said. Software comprises this framework, and it has to be built on open, standards-based technology. Standards-based computing with interoperability is critical for a number of reasons. It permits the user more choices; it enables end-to-end administration and security; it is necessary for multi-vendor, multi-platform computing; and it permits easier collaboration and integration of information sharing.

She said that when IBM began restructuring its own IT infrastructure in 1993, the company had to step back and define a consistent architecture for the corporation—a large challenge for a company of 350,000 people operating in 150 countries. It decided to move to a standards-based system. One step was to consolidate 155 data centers into the current number of seven. Another step was to replace 31 segregated networks by a single network. IT costs were cut by 25 percent, saving the company billions of dollars. She said that the federal government could achieve the same kinds of results, since it was already very dependent on open, standards-based infrastructure.

The second important policy she suggested was that government should be an early adapter of new technologies. "We've heard how important this is as we apply new technologies," she said. "But for some technologies, government may be the only entity with the size, the resources, and the mission to bring advances forward." She said this often meant bringing "true laboratory, detailed research"

---

[26]The advantage of a standards-based software infrastructure, as opposed to a proprietary infrastructure, is that it provides a common, publicly accessible platform that is freely available to multiple users. Proprietary systems usually make such functions as data transfer or sharing more difficult and expensive.

out of the lab and into the marketplace. An example she cited as important to her business was supercomputing. Without the a good partnership between industry and government, she said, it was unlikely that either partner could have produced the significant advances in commercial supercomputing "in this country or the world." She recommended that government leaders enable cutting edge technologies by smart investments that create new abilities for homeland defense.

Dr. Altman summarized her discussion of information overload by saying that integrated information architecture was the first and most important area to consider. Success depends largely on organizing information lines and applying a common information strategy across the missions of various agencies. "We all know," she said, "that disjointed information silos residing in many agencies will not get the job done. So government leaders need a free flow of information across their organizations and information systems that facilitate that."

The second area she discussed was partnerships. "It's a complex arena," she said, "and no one can go it alone." Partnerships need to draw on the combined expertise of government, business, academic institutions, and citizens. Each member of this partnership brings unique information and abilities for optimizing homeland defense decision making.

The third area was technology. "I think technology is key to cementing the partnership," she said. "We believe that it's the underpinning of open, standards-based architecture, allowing communication between various systems."

She said she had recently been asked how a business model translates into information technology. Ultimately, she said, business models translate into the software that is enabled through middleware, "and if we don't depend on that open, standards-based approach, ultimately these business models will fail."

Her last topic was public policy. "We should embrace sound public policies," she said, "that enhance our ability to exploit technology and achieve missions." This, she said, means intelligent investment in technology, research, and skills development. It means balancing the need to collect and share data with the need for appropriate security and privacy policy. And it means leadership—to promote interagency cooperation and to serve as an early adopter of critical technologies and standards.

NEW TECHNOLOGIES FOR NEW THREATS

Ronald M. Sega
Department of Defense

Dr. Sega began by saying that partnerships are very important to the Department of Defense (DoD), especially during its current transformation toward the new century. The current century, he said, is expected to be characterized by more uncertainty, and to demand a higher rate of technological change to prepare for that uncertainty.

The general direction of the DoD's transformation was defined in the Quadrennial Defense Review (QDR) submitted to Congress on September 30, 2001, which emphasized three cross-cutting initiatives: (1) national aerospace, (2) surveillance and knowledge, and (3) energy and power. The QDR included the following statement about transformation: "The evolution and deployment of combat capabilities provide revolutionary asymmetric advantages to our forces." The concept of asymmetry, he said, implies that "things are not the way they were in the twentieth century, as evidenced by the current global war on terrorism."[27]

The "critical capabilities" expected to bring about advantages were these:

Protect bases of operations

The DoD moved quickly after September 11 to combat terrorism, strengthen chemical and biological defenses, bolster missile defense, and plan for consequence management. On September 19, he said, only eight days after the terrorist attacks on the World Trade Center, the DoD formed a Combat and Terrorism Technology Task Force, bringing together leadership of the services and agencies. Two days later the task force had identified roughly 150 candidate technologies that could be brought to bear within approximately 30 days. A week or so after that it added additional participation from other agencies. On September 21, the group also decided to accelerate the development of several technologies, and was able to bring them into the field before the end of the year.

Conduct information operations

This included defensive information operations (IO) and information assurance. The strategy was to be able to communicate among all segments in the services and among all coalition partners.

Project and sustain U.S. forces

Top priorities were the detection of chemical, biological, radiological, and nuclear threats, as well as mines and other explosives.

Deny the enemy sanctuary

This involved a strategy of persistent surveillance, tracking, and rapid engagement with precision strikes. Tools included remote sensing and Enhanced

---

[27] The concept of asymmetric combat began to receive serious attention from the DoD in the mid-1990s, especially in light of the "asymmetric" tactics adopted by terrorists to circumvent traditional U.S. military strengths. The 1997 Quadrennial Defense Review, for instance, stated, "U.S. dominance in the conventional military arena may encourage adversaries to use . . . asymmetric means to attack our forces and interests overseas and Americans at home."

C4ISR (Command, Control, Communications, Computers, Intelligence, Surveillance, Reconnaissance). "We've been doing that for a while," said Dr. Sega, "and the theme is to know the space in which you're operating. That may be an urban warfare environment, including just a few city blocks; it may be a battle space; it may be the country." Denying sanctuary also makes use of unmanned aerial vehicles, long-range precision strikes, small-diameter munitions, and munitions to defeat hard and deeply buried targets. He added that the DoD had just created a new entity called the Northern Command, which was "responsible for the defense of North America."[28] The command plays a defensive role, and also supports civil authorities when requested for consequence management.

Conduct space operations

The primary goals of this mission are to ensure access to space, protect space assets, conduct space surveillance and overview, control space, and make use of sub-orbital space vehicles. An essential capability is to be able to cue other sensing systems.

Leverage information technologies

High-capacity, interoperable communications are essential to allow the Army, Navy, and Air Force to communicate with one another, said Dr. Sega. It is also important for coalition partners and the new Northern Command to communicate with civil authorities. The North American Air Defense Command in Cheyenne Mountain must be able to communicate with the air traffic control system controlled by the Federal Aviation Administration. "Technology can improve how we do that," he said, "and reduce the labor required."

Attributes of the new defensive strategy include knowledge, agility, speed, and lethality. Each, he said, is important for approaching the next century.

Major DoD Initiatives

The first initiative that takes advantage of these attributes is the National Aerospace Initiative. Its projects are organized under the headings of hypersonics,

---

[28]The U.S. Northern Command (NORTHCOM) is based at Peterson Air Force Base in Colorado Springs, CO. Its mission is not "homeland security," which is the objective of the new Department of Homeland Security, but "homeland defense." Homeland defense is defined as "the protection of U.S. territory, domestic population and critical infrastructure against military attacks emanating from outside the United States." In addition, NORTHCOM's mission includes "civil support to lead federal agencies," although the U.S. Posse Comitatus Act prohibits "direct military involvement in law enforcement activities."

space access, and space technology. A key partner in this initiative is NASA. An underlying objective of the initiative, he said, is to increase speed, eventually moving toward another generation of reusable launch vehicle. A goal is to reach roughly mach 12 by 2004 and to develop missiles and aircraft with access to space. This initiative will develop partners in the federal government and industry.

One example is a Hypersonic Flight Demonstration Program sponsored jointly by the Navy and DARPA. A 183-foot-long "scramjet" vehicle had been tested successfully at mach 6.5 at 100,000 feet nearly daily for several months, and since that time in a wind tunnel. The project was supported by technical direction from academia (Johns Hopkins University) and from industry (Boeing). "It's important to bring together resources from across government and into the private sector," he said.

A second initiative is Surveillance and Knowledge Systems, which includes sensors and unmanned vehicles, high-bandwidth communications, information and knowledge management systems, and cyber warfare. It has direct applicability to NORTHCOM, with some of its activities being performed by other agencies. He stressed the importance of biosensors; the DoD volunteered in the fall of 2001 to help in the effort to calculate the anthrax mortality curve, and to help the postal services calculate the right level of electronic, gamma, and x-radiation for irradiating letters and packages.

He also stressed the importance of high-bandwidth communications, and the "information assurance" that underlies cyber warfare. "We do have to bring together this information, and an understanding of our environment, so our knowl-

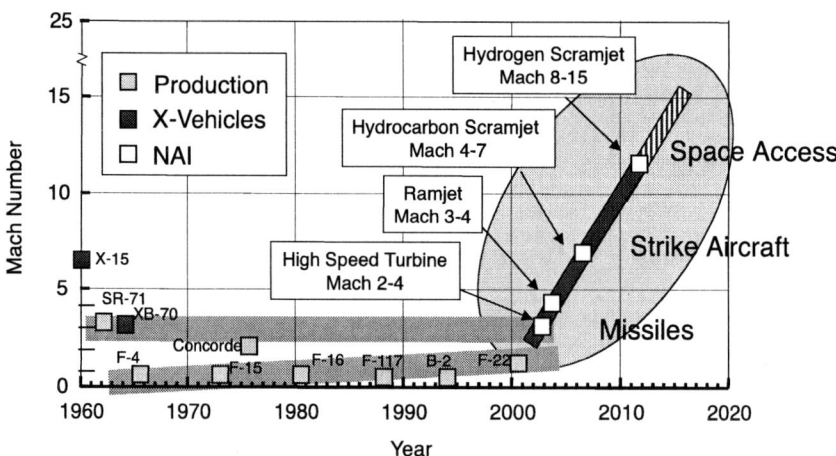

FIGURE 20 National aerospace initiative—Hypersonics.

edge systems will become increasingly important." These systems are now important down to the perspective of the single soldier in the field, he said, who is able to exchange information and digital resources with an operations center. Surveillance and knowledge has become a central part of understanding the warfare environment. A side benefit, he said, is the ability to gather additional information about atmospheric temperature, pressure, and humidity that can enhance weather forecasting, climatology, and other research.

He listed a number of "near-term transformation capabilities," including knowledge management, communications and networking, sensing, and information security. Examples included adaptive sensor webs, multilevel fusion, sense making—building cognitive bridges between understanding and deciding—decision making, modeling and simulation. These areas would all blend activities being done at universities with those at industry and other government agencies.

In the area of energy and power, the agency had looked at programs industry has launched and areas where government needs are complementary. On some areas, partnerships will be formed, and the DoD was then mapping that out.

Technology Transition is "Critically Important"

Technology transition he described as "critically important." It would be done in a variety of ways, depending on the complexity of systems and available partners. "Flexibility is one of the keys," he said, "whether it is a rapid method for just a few of these or a major acquisition program. You need different mechanisms to have effective transitions." It is important, he said, to emphasize systems engineering from the outset of each program, especially as systems become more complex. As an example, he said that at the September 21 meeting, the task force decided to accelerate development of a thermobaric rocket designed to strike targets hidden within hardened, deeply buried tunnel complexes.[29] "We were at a basic chemistry state in the Navy in October, doing laboratory tests, modeling, and simulation. By November were doing static tests in Nevada. On December 14 we had a flight test." A partnership consisting of the Navy, Air Force, Defense Threat Reduction Agency, DoD, and industry moved the program from basic science to a certified system in less than 90 days. "So you can go quickly," he said, "and in some areas we're trying to go faster than that."

He stepped back in perspective to discuss relative investment of the DoD in the U.S. and worldwide research base since World War II (see Figure 21). In 1965, he said, the DoD was supporting about 25 percent of all R&D, both U.S. and worldwide. At the current time, the DoD supports less than 5 percent of U.S.

---

[29]Some thermobaric devices are designed to penetrate doorway systems and even rock shielding before exploding inside tunnel complexes.

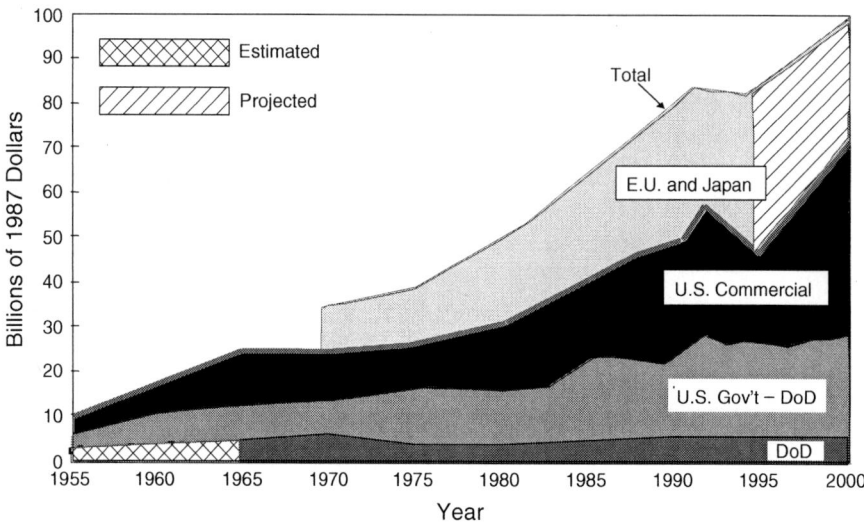

FIGURE 21 U.S. and worldwide research base since World War II. SOURCE: Report of the Defense Science Board Task Force on the Technology Capabilities of Non-DoD Providers, June 2000; Data provided by the Organization for Economic Cooperation and Development and National Science Foundation.

and worldwide research. "That means," he said, "that in the DoD we need to work with and take advantage of all the research that's being done outside the department. A lot is in the United States, some is outside. But much of that is also available to an adversary. So we need to make sure that every dollar we spend is highly leveraged, so we will in fact have a technological advantage if the time requires it."

In summary, he said, technology is the foundation for DoD objectives into the future. "We are aligning our programs with these general goals across services and agencies, and collaborating with others outside the DoD. We provide an integrated approach to the effort, and I believe that increasing partnerships is going to be key for national security." Combating terrorism with technology is also a model others are using, including NATO, with whom the U.S. has collaborated in applying technology against terrorism.

He concluded with a word about manpower, and the need for more trained scientists and engineers. "If we want to continue to innovate," he advised, "we're going to have to pay attention to the workforce. In that area we're all in this together."

SECURITY CHALLENGES IN AN OPEN ECONOMY

Steve Flynn
Council on Foreign Relations

Mr. Flynn started his lecture by saying the nation faces the "daunting challenge" of "transforming a security paradigm that was built for the federal government exclusively" into a partnership involving the rest of civil society. He mentioned that his lecture would cover three general points surrounding this much needed paradigm shift: (1) Such a partnership would not be possible unless the organizers convey a genuine sense of urgency. (2) There are "a few axioms" that might guide a sustainable private-public relationship with regard to security. (3) There is an example of how this might be accomplished by taking advantage of a great but under-appreciated strength of our society—the ability to draw on public-private relationships at the local and community levels.

After outlining his talk, Mr. Flynn commented that "America, a year later, is dangerously under-protected and unprepared for a catastrophic terrorist act." He suggested that the nation was in a more dangerous time, post-September 11, for three reasons.

(1) We have discovered that "David" can attack "Goliath"; that sheer force cannot always defend against asymmetric and unexpected tactics.

(2) Enemies of the United States might previously have been restrained by America's "superman aura," or "the perception that there was a CIA agent under every rock, a satellite over every head." September 11 demonstrated that whatever such buffer might have existed had been removed, and our civil society had been found to be "wide open." "We don't have protection built here," he said. "We built our civil society largely for efficiencies and for integration and all that goes with it, and we didn't put up much protection."

(3) The most compelling reason is that "most of the damage done post-September 11 is what we did to ourselves." As an illustration, he described our "reflexive response" in first grounding all civil aviation, and then closing our seaports and effectively sealing up the borders with Mexico and with Canada. "We did something no one else could do," said Mr. Flynn. "We imposed a blockade on our own economy as an effort to make ourselves more secure." We essentially tried to "freeze globalization" in order to sort out what had happened.

That mistake, he said, was quickly apparent. Within two days, after closing the Ambassador Bridge between Detroit and Ontario had created an 18-hour queue, Daimler Chrysler reported they would be closing an assembly plant on the following day; Ford Motor Co. announced that they would be closing five assembly plants during the following week.

Efficient Systems May Not Be Secure Systems

The global trading system, according to Mr. Flynn, was built to maximize "efficiency, low cost, and reliability." "It had no security built into it and the folks who were responsible for filtering the bad from the good basically stopped trying a long time ago." He suggested that the only tool at the disposal of the surface and maritime transportation systems is "a kill switch," the ability to shut down whole system.

He recalled that it had taken the nation three days to perform security inspections of all commercial aircraft after September 11. To complete an analogous inspection of all cargo containers in the United States would probably take more than six months. At any given time, there are some 16 million containers in use around the world. Because they cost only about $1500 apiece, they are relatively easy to purchase, to fill with cargo, and then, with minimal documentation, to deliver to any container port in the world. This system moves some 80 percent of the world's general cargo, he said. "It allows you to move 15 tons of material from Europe to the United States, on the average, or from Asia, for $900–1200."

This container system, he said, is what makes outsourcing possible: the ability to send work abroad and use highly compressed production cycles to quickly move products from the design stage to development and back to consumers in the U.S. or elsewhere. The resulting ability of companies to maintain "razor-thin" inventories has been a significant part of U.S. competitiveness over the last ten years, which capitalizes on this revolution in transportation.

This is, however, a system largely without security. Its architects built it for low cost, efficiency, and reliability. Security was viewed as something that worked against the system: raising costs, undermining efficiency, and undermining reliability. "For someone like myself," he said, "who has spent the last ten years talking about this with the architects of the container system, I was like the teetotaler at the New Year's Eve party. Now we are trying to retrofit security into this system. The approach we're taking is like trying to retrofit a split-level ranch house to make it handicapped-accessible. It's expensive, it's ugly and likely to not work very well."

"How can we develop the right approach?" he asked. The key, he suggested, was to realize that many things we do for single-point security in fact make society less secure. This is important to identify, he said, because it offers the opportunity to see how security and efficiency are not opposed, but can come together. He offered the example of security at the U.S.-Mexico border crossing in Laredo, Texas. "You could not design a better system for organized crime," said Mr. Flynn, "than we have at the secure border in Laredo."

The problem begins with the long delays for truck inspection. Since the owner of a quarter-million-dollar load cannot afford a typical six-hour delay, the market has developed its own practical solution. In a typical scenario, a long-haul truck approaches Laredo from the north and drops its trailer at a depot in North Laredo.

A short-haul truck comes to pick up the load and take it to a Mexican customs broker, who verifies its compliance with Mexican customs law. The short-haul truck then gets in line at the border, crosses some hours later, and drops the load at a depot on the Mexican side. There it is picked up by a Mexican long-haul driver and taken to the interior. The mean pay for short-haul drayage is close to $750 a load, whether it takes six hours or an hour and a half. These short-haul firms are typically mom-and-pop firms using worn-out tractors, and the fee is reasonable. The system itself, however, introduces a new degree of inefficiency, corruption, and opportunities to circumvent security.

As a result, he said, after "hardening" the border in response to September 11 and increasing inspection times, the border today is less policeable and less secure than it was ten years ago. A corollary to that conclusion is that the more efficient the system is, the more policeable it becomes. An efficient system brings confidence about where a load of cargo originated and the chain of custody it has followed through the system. There is a security rationale to address and redress the "keystone cop" type of regulatory oversight that plagues the current border system. At the same time, this rationale would now be resisted by the mom-and-pop employment that thrives on the current system.

He used the analogy of the national Interstate Highway system. Although the official rationale for the system was to provide an infrastructure for national defensive mobilization, the real benefits have been commercial rather than defensive. President Eisenhower, in promoting the system, had to overcome the same kind of local resistance from local communities and states that did not want to lose locally owned restaurants and the local road-building that was a major source of graft in every state.

A Secure System: Tracking and a "Chain of Custody"

For both trucks and containers, Mr. Flynn urged reforms that would both focus more sharply on the fundamental objective of security and do away with the kind of inefficiencies that had given rise to the short-haul trucks in Laredo. As an example, he reviewed how the movement of the 16 million containers might be done more securely. At its heart, he said, container security requires two elements: "First, can I have any confidence that what's been loaded into this conveyance is legitimate and authorized. And second, when it's on the move, do I have any confidence that it hasn't been intercepted or compromised."

If the answer to both those questions is yes, he said, there is no need to interrupt the commercial flow of containers and provoke the negative economic results that would surely follow. Instead, one goal of a more secure system would be to expedite the flow of containers. Goods are most vulnerable when they are at rest, he said. Another goal of a more secure system would be to create an effective public-private partnership. The tools needed to do that begin with information. The system needs to know as early as possible what cargo is moving toward your shores so you can assess its legitimacy. Next, it needs an auditing capability that provides confidence that the cargo is legitimate and authorized. Finally, it needs a chain of custody that provides confidence as it moves through the system that each link in the chain is in fact low risk.

How can such a system be built? he asked. It would require both sensors that reliably report on the integrity of the interior of the container, and a system of tracking. Tracking, he said, is critical for several reasons. Most important, the only way to act on information that the integrity of a cargo has been compromised is to know exactly where the cargo is going. At present, the information that a chemical weapon is bound for the United States in a container does not provide security because there is no tracking system that identifies the port to which the container is headed. Without such knowledge, the information could lead only to mass disruption. A complete itinerary is needed to make the intelligence actionable. Also, the information allows for forensic analysis, which would be critical in restoring confidence after an event.

Good Tracking Brings Commercial Benefits

A good tracking system would also bring large benefits to the private sector worlds of shipping and merchandising. The whole world of supply chains and international logistics would benefit from a more efficient system of transit visibility and accountability—at the same time it would allow for public-sector policing. Shippers could write tighter contractual agreements with transportation providers, manage more complicated outsourcing schemes, and design even tighter production cycles. A company that knows with certainty where its products are can maintain thinner inventories. A previous effort toward supply chain visibility was not able to gain momentum; a security rationale would probably help reignite that movement.

Mr. Flynn said he had been working for the past year on an initiative of this type called Operation Safe Commerce. He had begun by approaching a company called Osram Sylvania, which makes light bulbs in Slovakia and has a distribution center in New Hampshire. In the wake of September 11, Osram Sylvania was eager for ideas about making their supply line more secure, and the company agreed to act as a test bed for research on how to do this. They agreed to let Operation Safe Commerce put sensors on their products and serve as a "real supply chain for real R&D" across the world.

To oversee this research, Mr. Flynn invited the U.S. attorneys of New Hampshire and Vermont, the district commander of the Coast Guard, the regional director of U.S. Customs and Border Protection, and the regional director of the Immigration and Naturalization Service. The TISHWIG released $200,000 to perform a test run in May and June 2002. Mr. Flynn said that that amount equaled the total funds spent by the United States on container security during the past year, compared with $200 million a month on baggage screening.

It turned out that the technology design suffered from a lack of communication among the engineers who created it and law enforcement experts who judged it too visible, but as a proof of concept it was a success. The tracking device arrived without incident in New Hampshire, as scheduled.

Mr. Flynn said that the good results of this concept and of the preliminary experiment should draw the support of the private sector—especially if the alternative to a tracking system is the threat of a shutdown in some portion of the container system. He estimated that a shutdown would cost about a billion dollars a day for the first five days, and then begin to rise rapidly as the entire container system slowed to a halt.

He summarized by reiterating his support for security solutions based on the actual operational features of ports and border systems. While the search for solutions needs the support of Washington, he said, solutions are most likely to be found "where the rubber hits the road," by employees of the port authority, by U.S. attorneys, and by people working in the field. "Bring the R&D down to those levels and do it quickly," he advised. "By engaging those folks, I think we're going to get somewhere in the kind of hurry we need."

DISCUSSION

A questioner who had worked in the DoD noted that he had heard no discussion of the importance of a systems approach. He referred to the need for "a whole systems acquisition process," which includes mission analytic work and a technology-based acquisition strategy to solve problems. He cautioned that such an approach should not be left out of the thinking of the workshop, and advised against a strategy that was too narrowly "a fixation on a technology innovation agenda."

Dr. Sega said he agreed that "systems are critical here. We're talking about a network, with different systems, all of which have to interface effectively." He said that systems included people in the field, connecting back to numerous agencies and other people in the field, and IT systems, and systems development as you begin to tackle the task. "I couldn't agree more that a systems approach is going to be critical here, and I think Steve Flynn really was talking about systems."

Mr. Flynn readily agreed that "a single point approach can backfire on you from a security standpoint as well as an efficiency standpoint." The essence of Operation Safe Commerce, he said, was to look at how the real world works; it is intermodal and global. The most realistic way to do security gap analysis, he said, is to examine the entire system in question without preconception about where the real problems—or solutions—would be found. Some of them would be in paperwork, some in relationships, and some in the standards used by terminal operators. "We need all that data," he said.

The second stage, he said, would be to apply the appropriate R&D for addressing those gaps along the entire length of the container transportation system. He said that his real concern was that someone would examine only the individual parts of the system in isolation; "that's not how the real world works." He gave the example of developing sensors that might work in one part of the system but not in another: "Some are designed to work well in railroad cars, but don't work well in salt water; or they don't do well when they are dropped eight stories by a gantry crane." The key, he said, is that the research and development has to be applied to an actual supply chain, under real-life working conditions, in order to win the confidence of the commercial world and the public. "Otherwise, he said, "we haven't done anything but waste money. It's all about a systems approach."

Dr. Altman recalled the example of the Transportation Security Agency, where a systems approach had to be applied to the business vision of how TSA would operate. Unlike systems in the past that might have been closed systems, she said, this was an open system with connections to a variety of different sources of input and output.

A questioner reminded the participants that the data needed at decision making points in the system required program managers who are both "techno-literate in the information age" and able to have a "program management sense" of an acquisition strategy that uses cutting-edge, high-performance technologies.

The Need for Behavioral and Social Research

Another discussant reminded participants not to leave out the behavioral and social sciences in addressing terrorism. "Whole systems have people in them," he said, "and as Dr. Altman just said, the systems are open." He said that a bioterrorist incident would create public problems that have to be dealt with by behavioral and social scientists. First, how do you get people to abide voluntarily by quarantine rules? This is essential to prevent the spread of an infectious agent: How can people be prevented from panic and creating gridlock on evacuation roads that would block response equipment from reaching the site of the incident?

Second, he said, working within systems brought two challenges. One is how to bring different elements of technology together, and the other is how to bring disparate elements of the culture together. A area of social science called organization development develops skills in inducing people to work with technology and with each other. He urged his audience not to lose sight of the need to take both scientific and human systems approaches to these problems.

Mr. Flynn expressed strong agreement. He said that a challenge in seeking funding for the first phase of Operation Safe Commerce was that no one wanted to pay for the human part of the process. "Everybody would pay for a new sensor, a new sniffer, gadgetry to hand around, but they didn't want to pay for convening the meeting among Customs, Coast Guard, INS, and attorneys general."

Another issue, he went on, was the tendency to use all manners of technology that may or may not be appropriate, such as sensors designed for sanitized facilities. He cited an incident that startled security forces and much of the country in September 2002 when a container ship approaching New York harbor was seized on the basis of misinterpreted radiation readings. Inspectors had not been properly trained, and port personnel had not been sufficiently briefed to deal with such an alert, and the public was unsettled by rumors for many days.[30]

"Ports are complicated urban environments," said Mr. Flynn. "If we use technology developed in isolation from that world, it won't work, and the agents lose all confidence in the devices. People haven't been trained to use them and there's no backup support when something goes wrong. These kinds of training investments must be made, and hopefully someone from behavioral and social sciences will point that out before we throw around more technology and cause more mischief."

Dr. Altman said that many lessons could be learned from private enterprise about social and cultural adaptation to change. Certainly, she said, an imperative like an impending or existing terrorist event creates the impetus for change. In industry, she said, "we have learned through leadership and very specific approaches to change the culture rapidly in such a way that many people, even across companies, embrace change very rapidly and accept it as a means of improving the ability to execute." She urged a careful look at such lessons.

Better Training for First Responders

A discussant referred to a recent report by the Federation of American Scientists on how the nation might improve training for those likely to face the consequences of terrorist acts.[31] To begin with, he said, first responders would have to learn new behaviors of reaction and response to terrorist threats. It was well known to behavioral scientists, he said, that typical training programs anticipate a

---

[30] Two days before the first anniversary of September 11, the Liberian-flagged ship *Palermo Senator*, carrying 655 containers, was stopped entering New York harbor by a Coast Guard inspection team. When trace amounts of radiation were detected from the cargo, multiple agencies were summoned and the ship was ordered six miles out to sea where it was detained for three days. The cause turned out to be background radiation from ceramic tiles, detected by a sensor that was not designed for shipboard use. That particular sensor, for example, could be triggered by the potassium nitrate emitted by a cargo of bananas.

[31] Henry Kelly et al., "Training Technology Against Terror: Using Advanced Technology to Prepare America's Emergency Medical Personnel and First Responders for a Weapon of Mass Destruction Attack," Federation of American Scientists, September 2002. The Introduction states, "Without an effective investment in training, the nation's investment in weapons of mass destruction (WMD) response will be largely wasted."

memory retention of six months or less, and in an effort to increase this, military and other scientists were beginning to learn more about how people learn. This knowledge could be used to develop more efficient and effective training for the military and response communities. "We've managed to wire all our schools," said the discussant, "but we have almost no educational content transmitted through those wires—much less sophisticated content that truly enhances the learning process." He said that the Federation of American Scientists was working with NORTHCOM and the National Guard, among others, to develop sophisticated training content that would take advantage of new learning techniques from the behavioral sciences.

Mr. Flynn said he had been asking himself how much security is truly needed. He said that no matter how much we invest in a safer homeland, there would inevitably be "incidents," because the United States is an open society and "we're in a dangerous world, and there are people out there who are intent on causing harm." But should there be an incident, he said, an essential task is to determine whether it was a result of a correctible breach in security or of an absence of security. "If people view it as a breach in security," he said, "there will be a strategic pause, and then we'll get back to life as we sort it out. If they view it as the absence of security, they'll want to shut down the system until security can be put in place."

He said that the "military value" of catastrophic terrorism comes not from "killing people atop a landmark" but from the "profound disruption created by the incidents themselves." With an appropriate level of security, there is a low risk of mass disruption, he said, because people are willing to go through the "strategic pause" instead of "shutting down the system" until security is put in place. With this kind of behavior, he said, "you have deterrence." A terrorist would say, "Why would I bother to commit terrorism if it would simply accomplish a mass murder or vandalism with no tangible impact on U.S. power?" "This is something that our adversaries would consider before they commit these horrific acts," he said.

He called this "as much the guts of a counter-terrorism strategy as going to the source of al Queda." The key, he said, is that security is like safety; it is not an end in itself. Security is found in sustainable systems, such as global networks of trade, finance, labor, information, and transportation—that continue to function even in a world that contains people with malicious intent who want to disrupt those systems. The objective is to build enough safeguards and resiliency into the system to sustain it even during or after attack.

A second part of that objective, he said, is to think of safeguards and resiliency throughout the international extent of such systems. "If we keep this narrowly focused on the homeland," he said, "it's a bit like hiring a network security manager who says I'm just going to protect the server next to my desk; the others are too far away." Even though maintaining systems in an international context is far more complex, he said, it must be done.

Panel V

Roundtable on Partnering for National Missions: Defense, Health, Energy

INTRODUCTION

Patrick Windham
Windham Consulting

Mr. Windham offered several observations on the day's discussions. A theme that had run through the entire four-year STEP project, he said, was that in many situations government-industry partnerships had been shown to work. He agreed with Bill Spencer that there are necessary preconditions to success, and that there were no guarantees but rather a set of best practices.

He also said that in the case of homeland security, he had heard that partnerships were not only an option, but for many purposes a requirement. This was primarily because no single agency of the federal government—even the Department of Defense—had all the necessary technology to combat terrorism alone. Nor did sufficient capability exist across the government; homeland security would require the full participation of many people in the private and academic sectors as well, to implement appropriate technologies and responses.

He then posed a series of questions to the panel:

- What have we learned from other agencies about how to structure programs, and how to structure this new agency?
- How much should the Department of Homeland Security rely on its own R&D and how much on other programs?

- Under what conditions would high-tech companies want to work with the government? Clearly, he said, there is a patriotic intent to participate in programs that advance security. But commercial high-tech companies—even those with the best intentions and will to help—cannot afford to allow their business plan to be derailed. The challenge is to devise a working relationship that has benefits for all partners.
- In designing technologies and systems to work in ports, hospitals, borders, and other "real-life" environments, how can we integrate the experience of first responders in thousands of local communities—firefighters, emergency medical people, police? These people are unlikely to have high-tech backgrounds, but they know how large systems work, and their knowledge is essential to adapting and operating high-tech tools in the real world.

Christina Gabriel
Carnegie Mellon University

Dr. Gabriel began by noting the importance of a portfolio of approaches. Every new technology, every platform, and every sector has different qualities, and the only way to create an entity broad enough to comprehend these qualities is to create a partnership with diverse representation. Also needed is a variety of programs and approaches, both from social science and "real people."

"We must keep reminding ourselves," she said, "that one of the reasons we do technology is because we know how to do it. Technology has always been the route for the best and brightest people to get really exciting work." At the same time, she said, it may actually be easier to work on a technology problem than to address many of the critical but broader challenges facing the world "You hear that if someone were to come back from Biblical times," she said, "they would be astounded by the technology and wouldn't understand a thing about it. But if you told them about our current geopolitical tensions, they'd understand them perfectly." The point, she said, is that we have made too little progress in solving many of the complex social problems that are limiting the progress of people and nations toward a better quality of life and "the pursuit of happiness," which is why there are still "David's" out there trying to kill the "Goliaths." How do we make progress on those underlying issues?

In setting up any program, she said, it is important to have the policy agenda, the goals and objectives, the evaluation mechanism, and the leadership. But it is equally or more important to ensure that the program has the right operational features; specifically, what incentives invite the people naturally to work toward the goals of the program. When a group of agencies is forced to make joint decisions about funding allocations, for example, the group has to work in collaboration or nothing will be accomplished. She also advised that the group have diverse expertise, not only in the specific sector under study, but also in related sectors. As an example, she suggested that people who work in government after their academic training and a career in industry are often valuable candidates because of their insights from multiple sectors.

William Spencer
International SEMATECH

Dr. Spencer first offered a simple definition of a partnership as "two or more entities that get together to do something." Then he looked more deeply into the essence of one kind of partnership to suggest a critical function. This was the partnership that led to the development of thermal ink-jet printers. Lord Kelvin had first suggested the idea, and even patented it, and Siemens actually built the first ink-jet printer in 1951. But the real story, he said, depended on two companies, one U.S. and one Japanese, which simultaneously invented a means to build ink-jet printers and a way to make them commercially feasible.

The two companies, Canon and Hewlett-Packard, made their inventions independently, and "the usual approach," he said, "would have been to hire lawyers and go to court." Instead, said Dr. Spencer, who was then working at Xerox PARC "across the street from H-P," the CEO of H-P and the CEO of Canon agreed that they would independently pursue the technology in the market, in both cooperation and competition. Today, thermal ink-jet printers outsell laser printers by 12 to one, and the ink-jet cartridge business has revenues of $20 billion per year. "There's not a lot of profit for either company in building printers," he said, "but there is a lot of profit in selling cartridges." He estimated that 50 percent of the profits of both companies came from cartridges.

In summary, he said, this is what a partnership is like—Canon and Hewlett-Packard sharing technology but competing in the market, and together building an entire market that had not existed. "If we're going to have partnerships in homeland defense," he said, "and I believe it's a really good idea, government and industry have got to realize what a partnership is. I would suggest that studying the HP-Canon partnership is a good idea, along with the publications we've put out in this study."

William Bonvillian
Officer of Senator Lieberman

Mr. Bonvillian said that the panels had concentrated mostly on the domestic aspects of homeland security, and reminded the groups to think of terrorism as an international problem. Creating a department in the United States, he said, and creating new defensive elements in this country "only scratches the surface." The country has to learn "how, in effect, to push out our borders and our international connections in a way that I don't think we've spent much time thinking about, at least in the defense sector." He referred to the kind of system suggested by Steve Flynn, and the observation that unless Customs and the Coast Guard have "a good fix" on what is being shipped to our shores by a Czech light bulb manufacturer, "we will not have a secure system at home." He said that another partnership challenge is to imagine and build new international partnerships and relationships that come together as a workable and effective system.

James Turner
House Science Committee

Mr. Turner said that DARPA, created in 1957, was the last good model he would recommend for building an R&D program within a cabinet department such as the one contemplated now. A model he did not recommend for structuring the Department of Homeland Security was that of the Department of Energy, created in 1977 in response to the Arab oil boycott. There were some striking similarities to the present: The country was reacting to a crisis, and the justification was national security. The solution then was to merge parts of disparate agencies and also to create new "boxes" for other functions.[32] The new department had the traditional structure of a secretary on top, with an undersecretary supervising all R&D, just like the DHS. There was also, he said, an "unnecessary amount of diversity in the agency," caused by overbroad legislation that "threw a lot of disparate problems together" including the regulatory Federal Energy Administration, the Energy Research and Development Agency, and the Federal government's nuclear weapons program. Major related components were left outside the DoE—in the Environmental Protection Agency, the DoD, NASA, and so on, making complex coordination necessary to merge bureaucracies.

The first big issue for DHS, he said, was how the government employee unions were treated. In creating the DoE, streamlining did not occur; instead, the field structures of what had been in the FEA and ERDA were both kept. "Both structures are still in place today," he said, "and still conflicting with each other, in my view." DoE even had an advantage, he said—a 20-year history with ERDA, and the earlier Atomic Energy Commission, during which its R&D programs were unified. DHS, by contrast, would have to absorb research pieces of CDC, NIH, and other agencies and create a new R&D structure.

"Two things happened with DoE," he said. "From day one, the top management couldn't think about R&D; they had to think about the crisis of the day. That still happens today, and I expect that it is going to happen with this agency." Secondly, he said, the DoE did not achieve the objective of weaning America from dependence on foreign oil. When the DoE was created, the United States was importing less than 50 percent of its oil; today it imports 58 percent of its oil. "So the point is that we created a wonderful agency on paper. That doesn't mean it will achieve its goals."

---

[32] An example was the Office of Commercialization within DoE, which disappeared soon after its creation.

There were differences today, however, he said, which was "good news." The nation had had 25 year of experience with partnerships, some of which had worked well (SEMATECH) and others that had not worked so well (Synfuels). There was also the benefit of a history of reviews and analysis of what had worked for SEMATECH, the ATP, and some NIH programs. Research on university-industry relations, including work by Dr. Gabriel, he said, had helped us understand more about how partnerships work.

He concluded by saying that "the default position is going to be failure." Unless the DHS was designed with the lessons of the past in mind, "R&D is going to be buried by the crisis of the day. We have to design DHS in such a way that that doesn't happen." Organizationally, said Mr. Turner, the DHS would be successful only if it is organized as a single department with a common mission, not as a holding company for a variety of agencies with piecemeal missions. He said that the achievements of the first secretary and the first undersecretary would be critical, as would the organization's executive orders, which is where the executive branch would receive its guidance on objectives and implementation.

He reminded the workshop that "homeland security" is "an important value but not the only value; we have to design with civil liberties in mind, the environment, and the other values that make this country great."

He also suggested that the Academies had a more important role in creating DHS than they had had during creation of the Department of Energy 25 years earlier, partly because the Office of Technology Assessment and several other objective sources of information no longer existed. "This has been a great meeting," he concluded, "but the work of the Academies is just beginning. The Academies are our best convener of experts right now, and I think they will be the best constructive critic to keep this department on the right track."

Closing Remarks

Gordon Moore
Intel Corporation

Dr. Moore reiterated that STEP had been looking at public-private partnerships for more than four years, and that some of them had worked well. He thanked some of the people who had made that four-year survey possible, including the key financial supporters, the NRC staff, headed by Chuck Wessner, the STEP board itself, and the GIP Committee, especially his vice-chair Bill Spencer.

He asked to remind the workshop of one perspective that should not be forgotten, which are "some deep partnerships between government and industry that are implicit rather than explicit." These kinds of partnerships, he said, were responsible for such achievements as creating an environment in which innovation can take place and be exploited. They had also promoted education and training, regulation, and the laws that govern how organizations behave, such as anti-trust laws and intellectual property laws. In addition, the structures of taxation, fiscal policy, and monetary policy also formed a kind of partnership that had made the United States the most productive place in the world to create technological innovations and transfer their value to the marketplace.

He closed the workshop by thanking all panel participants for their presentations, and for demonstrating the ongoing importance of partnerships to the complex public issues of the day. "It's nice to see," he said in conclusion, "that our work is applicable to this major new problem area that the nation is forced to consider."

III

APPENDIXES

Appendix A

Biographies of Speakers*

BRUCE ALBERTS

Bruce Alberts is president of the National Academy of Sciences and chair of the National Research Council, the principal operating arm of the National Academies of Sciences and Engineering. He is a respected biochemist recognized for his work in both biochemistry and molecular biology and is known particularly for his extensive molecular analyses of the protein complexes that allow chromosomes to be replicated.

Alberts joined the faculty of Princeton University in 1966 and after ten years moved to the medical school of the University of California, San Francisco. In 1980, he was awarded an American Cancer Society lifetime research professorship. In 1985, he was named chair of the UCSF department of biochemistry and biophysics.

Alberts is one of the principal authors of *The Molecular Biology of the Cell*, now in its third edition, considered the leading advanced textbook in this field and used widely in U.S. colleges and universities. His most recent text, *Essential Cell Biology*, is intended to present this subject matter to a wider audience. He is committed to the improvement of science education; he helped to create City Science, a program for improving science teaching in San Francisco elementary schools.

---

*As of October 2002.

ANNE K. ALTMAN

Anne Altman is IBM's Managing Director for the relationship between IBM and the U.S. Federal Government. She has full business management responsibility for all aspects of IBM's 80+ year relationship with the Federal Government. Ms. Altman leads a team of several thousand professionals around the globe, wherever IBM and the U.S. Government work in partnership on information technology solutions.

Prior to this appointment in January 2001, she was Vice President, U.S. Federal Government, with responsibility for the sale of hardware, software and services to the federal market. In 1999, Ms. Altman was named Director of Marketing, IBM Global Government Industry. In this role, she was responsible for the development and deployment of business plans and marketing programs for local, regional, and central government customers around the globe.

Prior to 1998, she held a number of executive and management positions, leading IBM's worldwide software accounts managers, and positions in worldwide software operations, networking and software sales, and business development. Ms. Altman joined IBM in 1981 as a systems engineer working with the Federal Bureau of Investigation and the Department of Justice, and spent the next 12 years in sales and sales management roles for IBM working with federal government partners.

Ms. Altman currently serves on the boards of the Government Electronic Industry Association and the Armed Forces Communication and Electronics Association. She is an active participant in a number of government-related review boards, including the National Academy of Public Administration's Information Technology Compensation Committee that examined issues related to the retention of IT skills in government. Ms. Altman's recent work has been focused on the rapid integration of disparate technology-based systems, and secure, effective collaboration across organizational boundaries. She routinely delivers congressional testimony for IBM on topics concerning e-government and national security. She has assumed civic leadership responsibilities as co-chair for the annual Juvenile Diabetes Research Foundation fundraiser sponsored by the Federation of Information Processing Councils and the Industry Advisory Council.

Throughout her career, Ms. Altman has received numerous awards, including Federal Computer Week's 2001 Federal 100 Award, presented to leaders who have made a difference in federal information technology, and the CIO Council's 2002 Azimuth Award for outstanding IT service to federal officials. She has also written several opinion pieces in both national and trade publications, most notably *The Washington Times* and *The Public Manager*, on issues surrounding national security, e-government, and human resources trends in the federal government.

KATHY BEHRENS

Kathy Behrens joined Robertson Stephens Investment Management medical group in 1983, becoming a general partner in 1986 and a managing director in January 1993. As Robertson Stephens Investment Management's first biotechnology analyst, she expanded the firm's health care presence by moving into emerging medical technologies. After nine years in research, Dr. Behrens joined the Venture Capital Group in 1988 and has since founded three biotechnology companies: Protein Design Labs, Inc., COR Therapeutics, Inc., and Mercator Genetics, Inc. She has been instrumental in raising over $1 billion in the public and private markets for biotechnology companies.

Prior to joining Robertson Stephens Investment Management, Dr. Behrens was a biotechnology analyst at Sutro & Co., Inc. She is a director of Abgenix, Inc. and Oncology.com and represents the interests of RS Investment Management in Mitotix, Inc., Onyx Pharmaceuticals, Inc., and Tularik, Inc. In addition, Dr. Behrens held board seats at Protein Design Labs, Inc., from 1986 to 1992, Cell Genesys, Inc. from 1990 to 1996, InSite Vision, Inc. from 1990 to 1995, COR Therapeutics, Inc. from 1988 to 1995 and Mercator Genetics, Inc. from 1993 to 1997.

Dr. Behrens has been a director of the National Venture Capital Association since 1993 and was President of the NVCA from May 1998 through April 1999. She served as Chairman of the National Venture Capital Association through September 1999 and is currently Immediate Past Chairman. She holds a Ph.D. in microbiology from the University of California, Davis, where she performed genetic research for 6 years.

ARDEN BEMENT

Arden L. Bement, Jr., was sworn in as the 12th director of NIST on December 7, 2001. Bement oversees an agency with an annual budget of about $819 million and an onsite research and administrative staff of about 3,000, complemented by a NIST-sponsored network of 2,000 locally managed manufacturing and business specialists serving smaller manufacturers across the United States. Prior to his appointment as NIST director, Bement served as the David A. Ross Distinguished Professor of Nuclear Engineering and head of the School of Nuclear Engineering at Purdue University. He has held appointments at Purdue University in the schools of Nuclear Engineering, Materials Engineering, and Electrical and Computer Engineering, as well as a courtesy appointment in the Krannert School of Management. He was director of the Midwest Superconductivity Consortium and the Consortium for the Intelligent Management of the Electrical Power Grid.

Bement came to his position as NIST director well versed in the workings of the agency, having previously served as head of the Visiting Committee on Advanced Technology, the agency's primary private-sector policy adviser; as head of the advisory committee for NIST's Advanced Technology Program; and on the Board of Overseers for the Malcolm Baldrige National Quality Award.

Bement joined the Purdue faculty in 1992 after a 39-year career in industry, government, and academia. These positions included: vice president of technical resources and of science and technology for TRW Inc. (1980–1992); deputy under secretary of defense for research and engineering (1979–1980); director, Office of Materials Science, DARPA (1976–1979); professor of nuclear materials, MIT (1970–1976); manager, Fuels and Materials Department and the Metallurgy Research Department, Battelle Northwest Laboratories (1965–1970); and senior research associate, General Electric Co. (1954–1965).

Along with his NIST advisory roles, Bement served as a member of the U.S. National Science Board, the governing board for the National Science Foundation, from 1989 to 1995. He also chaired the Commission for Engineering and Technical Studies and the National Materials Advisory Board of the National Research Council; was a member of the Space Station Utilization Advisory Subcommittee and the Commercialization and Technology Advisory Committee for NASA; and consulted for the Department of Energy's Argonne National Laboratory and Idaho Nuclear Energy and Environmental Laboratory.

Dr. Bement has been a director of Keithley Instruments Inc. and the Lord Corp. and was a member of the Science and Technology Advisory Committee for the Howmet Corp. (a division of ALCOA). He holds an engineer of metallurgy degree from the Colorado School of Mines, a master's degree in metallurgical engineering from the University of Idaho, a doctorate degree in metallurgical engineering from the University of Michigan, an honorary doctorate degree in engineering from Cleveland State University, and an honorary doctorate degree in science from Case Western Reserve University. He is a member of the U.S. National Academy of Engineering.

SHERWOOD BOEHLERT

Utica native Sherwood L. Boehlert (R-NY), Chairman of the House Science Committee, was first elected to the House of Representatives in November 1982. He is currently serving in his tenth consecutive term representing Central New York. In the 2000 election, he again won all nine counties and received a convincing 60 percent of the vote in a three-way race.

Boehlert has served on the Science Committee since 1983, and was elected Chairman in January 2001. The Committee has jurisdiction over all federal nonmilitary scientific and technology research and development programs, on which the federal government spends more than $30 billion a year. The Committee has jurisdiction over NASA, the National Science Foundation, and research and development initiatives within the Environmental Protection Agency, the Department of Energy, and the Department of Commerce. In addition, the Committee has jurisdiction over civil aviation research and development and marine research.

Boehlert is the third-ranking member of the House Transportation and Infrastructure Committee, serving as Chairman of its Subcommittee on Water Resources and Environment from 1995 to 2000. He remains an active member of that Subcommittee. Boehlert also sits on the Subcommittee on Highways and Transit, and the Subcommittee on Railroads.

Boehlert was reappointed by House Speaker J. Dennis Hastert as a member of the Select Committee on Intelligence, where he is on the front line of important intelligence decisions faced by Congress. Boehlert is a delegate to the NATO Parliamentary Assembly, also at the appointment of the Speaker, where he serves as chairman of the Assembly's Scientific and Technology Committee.

Born on September 28, 1936, in Utica, New York, Boehlert is a graduate of Whitesboro Central High School and Utica College (Bachelor of Science, 1961). Before serving as Oneida County Executive (1979–1983), he was manager of public relations at Wyandotte Chemical (1961–1964) and served two years in the U.S. Army (1956–1958).

Boehlert served as chief of staff for two area Congressmen, Alexander Pirnie (1964–1972) and Donald Mitchell (1973–1979), where he became intimately familiar with the people, places, and issues of the 23rd District. In honor of his former boss, Boehlert was able to secure passage of legislation in 2000 to rename the Veterans' Outpatient Clinic in Rome as the "Donald J. Mitchell Department of Veterans Affairs Outpatient Clinic."

An avid New York Yankees fan and movie buff, Boehlert and his wife, Marianne (Willey) Boehlert, make their home in New Hartford, New York. They have four grown children and five grandchildren. When Congress is in session, he returns home each weekend to stay in touch with people he feels fortunate to represent in Washington.

WILLIAM B. BONVILLIAN

William Bonvillian is the Legislative Director and Chief Counsel to Senator Joseph I. Lieberman (D-CT). Prior to his work on Capital Hill, he was a partner at both the law firms of Jenner & Block as well as Brown & Roady. Early in his career, he served as the Deputy Assistant Secretary and Director of Congressional Affairs at the Department of Transportation.

His recent articles include, "Organizing Science and Technology for Homeland Security," in *Issues in Science and Technology* and "Science at a Crossroads," published in *Technology in Society* this past February. His current legislative efforts at Senator Lieberman's office include science education, homeland research and development, and nanotechnology legislation.

Mr. Bonvillian is married to Janis Ann Sposato and has two children. He received his B.A. from Columbia University; his M.A.R. from Yale University; and his J.D. from Columbia Law School where he also served on the Board of Editors for the *Columbia Law Review*. He is a member of the Connecticut Bar, the District of Columbia Bar, and the U.S. Supreme Court Bar.

MICHAEL BORRUS

Michael Borrus is a Managing Director of Petkevich Group, an investment bank focused on the health-care and information technology industries. Before joining the Petkevich Group, Mr. Borrus was a Co-Director of the Berkeley Roundtable on the International Economy (BRIE) at the University of California at Berkeley and Adjunct Professor in the College of Engineering, where he teaches Management and Technology.

He is the author of two books and over 60 chapters, articles, and monographs on a variety of topics including high-technology competition, international trade and investment, and the impact of new technologies on industry and society. For the last decade, he has served as consultant to a variety of governments and firms in the U.S., Asia, and Europe on policy and business strategy for international competition in high-technology industries. Mr. Borrus is a graduate of Harvard Law School and a member of the California State Bar.

GAIL CASSELL

Gail Cassell is currently Vice President of Infectious Diseases, Eli Lilly and Company. She was previously the Charles H. McCauley Professor and Chairman of the Department of Microbiology at the University of Alabama Schools of Medicine and Dentistry at Birmingham, a department that ranked first in research funding from the National Institutes of Health since 1989 during her leadership.

She is a current member of the Director's Advisory Committee of the National Centers for Disease Control and Prevention. She is a past President of the American Society for Microbiology, a former member of the National Institutes of Health Director's Advisory Committee, and a former member of the Advisory Council of the National Institute of Allergy and Infectious Diseases of NIH. Dr. Cassell served 8 years on the Bacteriology-Mycology 2 Study Section and as Chair for 3 years. She also was previously chair of the Board of Scientific Councilors of the Center for Infectious Diseases, CDC.

Dr. Cassell has been intimately involved in establishment of science policy and legislation related to biomedical research and public health. She is the chairman of the Public and Scientific Affairs Board of the American Society for Microbiology; a member of the Institute of Medicine of the National Academy of Sciences; has served as an advisor on infectious diseases and indirect costs of research to the White House Office of Science and Technology Policy, and has been an invited participant in numerous Congressional hearings and briefings related to infectious diseases, antimicrobial resistance, and biomedical research. She has served on several editorial boards of scientific journals and has authored over 250 articles and book chapters. Dr. Cassell has received several national and international awards and an honorary degree for her research in infectious diseases.

MARYANN FELDMAN

Maryann Feldman is currently the Policy Director at the Johns Hopkins University Institute for Information Security (JHUISI) of the Whiting School of Engineering. In addition, she is a Research Scientist for the Program on Entrepreneurship and Management in the Department of Mathematical Sciences and adjunct Associate Professor in the Department of Economics at Johns Hopkins University. Before beginning her work as Policy Director, Dr. Feldman was Research Scientist for the Institute for Policy Studies at Johns Hopkins University. Prior to her work at John Hopkins, she was Visiting Assistant Professor at the H. J. Heinz III School of Public Policy and Management, Carnegie Mellon University and Assistant Professor of Management and Economics at Goucher College in Baltimore, Maryland. As of January 2003, she will be Associate Professor of Business Economics at the Rotman School of Business at the University of Toronto.

Dr. Feldman is the author of over 40 referred articles on a variety of topics related to science and technology policy including the economics of science and technology, the location of innovative activity, and university technology transfer activities. Her research has been funded by the National Science Foundation, the Andrew W. Mellon Foundation, and the Advanced Technology Program.

Throughout her career, Dr. Feldman has received numerous fellowships and professional awards. She received a B.A. in Economics and Geography from Ohio State University, a M.S. in Management and Policy Analysis, and a Ph.D. in Economics and Management from Carnegie Mellon University.

KENNETH FLAMM

Kenneth Flamm is the Dean Rusk Professor of International Affairs at the LBJ School at the University of Texas–Austin. Before this, he worked at the Brookings Institution in Washington, where he served 11 years as a Senior Fellow in the Foreign Policy Studies Program. He is a 1973 honors graduate of Stanford University and received a Ph.D. in economics from M.I.T. in 1979. From 1993 to 1995, Dr. Flamm served as Principal Deputy Assistant Secretary of Defense for Economic Security and Special Assistant to the Deputy Secretary of Defense for Dual Use Technology Policy. He was awarded the Department's Distinguished Public Service Medal by Defense Secretary William J. Perry in 1995 as well.

Dr. Flamm has been a professor of economics at the Instituto Tecnológico de México in Mexico City, the University of Massachusetts, and the George Washington University. He has also been an adviser to the Director General of Income Policy in the Mexican Ministry of Finance and a consultant to the Organization for Economic Cooperation and Development, the World Bank, the National Academy of Sciences, the Latin American Economic System, the U.S. Department of Defense, the U.S. Department of Justice, the U.S. Agency for International Development, and the Office of Technology Assessment of the U.S. Congress. He has played an active role in the National Academies of Sciences' committee on Government-Industry Partnerships, under the direction of Gordon Moore, and played a key role in that study's review of the SBIR program at the Department of Defense.

Dr. Flamm has made major contributions to our understanding of the growth of the electronics industry, with a particular focus on the development of the computer and the U.S. semiconductor industry. He is currently working on an analytical study of the post-Cold War defense industrial base and has expert knowledge of international trade and the high technology industry issues.

STEVE FLYNN

Stephen Flynn is a Senior Fellow with the National Security Studies Program at the Council on Foreign Relations, headquartered in New York City. He is also a Commander in the U.S. Coast Guard, and a member of the permanent commissioned teaching staff at the U.S. Coast Guard Academy in New London, Connecticut.

Currently at the Council, Commander Flynn is directing a multi-year project on "Protecting the Homeland: Rethinking the Role of Border Controls." He is author of several book chapters and articles on homeland security, border control, transportation security, and the illicit drug trade. His recent publications include, "America the Vulnerable," in *Foreign Affairs* (Jan/Feb 2002), "The Unguarded Homeland" in *How Did This Happen? Terrorism and the New War,* PublicAffairs Books (Nov 2001); and "Beyond Border Control" *Foreign Affairs* (Nov/Dec 2000).

He served in the White House Military Office during the George H. W. Bush administration and as a director for Global Issues on the National Security Council staff during the Clinton administration. From August 2000 to February 2001, he served as a consultant on the homeland security issue to the U.S. Commission on National Security (Hart-Rudman Commission). He was a Guest Scholar in the Foreign Policy Studies Program at the Brookings Institution from 1991 to 1992, and from 1993 to 1994 he was an Annenberg Scholar-in-Residence at the University of Pennsylvania.

A 1982 graduate of the U.S. Coast Guard Academy, Commander Flynn received the M.A.L.D. and Ph.D. degrees in International Politics from the Fletcher

School of Law and Diplomacy, Tufts University, in 1990 and 1991. He has received academic prizes for his undergraduate and graduate studies. In 1991 he became the first Coast Guard officer to be selected as a Council on Foreign Relations' International Affairs Fellow.

Commander Flynn has lectured around the United States and abroad on the homeland security, border control, drugs, and crime issue, has provided testimony on Capitol Hill and before the Canadian House of Commons, and has served as a guest commentator for ABC with Peter Jennings, the Charlie Rose show, 60 Minutes II, CNN, National Public Radio, and BBC Radio.

Commander Flynn's afloat assignments include two tours as commanding officer of the Coast Guard Cutters REDWOOD and POINT ARENA, and one tour as operations officer of the Coast Guard Cutter SPAR. His professional awards include the Legion of Merit, Meritorious Service Medal, the Coast Guard Commendation Medal, and the Coast Guard Achievement Medal. In 1999, he received the Coast Guard Academy's Distinguished Alumni Achievement award.

CHRISTINA GABRIEL

Christina Gabriel is Vice Provost for Corporate Partnerships and Technology Development at Carnegie Mellon University. Dr. Gabriel comes to Carnegie Mellon from CASurgica, Inc., a Carnegie Mellon spin-off company focusing on computer-assisted orthopedic surgery, where she was President and CEO. In earlier university positions, Dr. Gabriel has served as Director of Collaborative Initiatives at Carnegie Mellon as well as Vice President for Research and Technology Transfer at Case Western Reserve University in Cleveland, Ohio.

Dr. Gabriel spent five years with the National Science Foundation in Washington, D.C., and Arlington, VA, most recently serving as Deputy Assistant Director for Engineering, which is the chief operating officer of the Engineering Directorate, an organization of 140 staff members (half PhD-level) that awards over $300 million to universities and small businesses for engineering research and education. In earlier assignments at NSF, Dr. Gabriel served as program director within several engineering research programs, as well as Coordinator for the $50 million university-industry collaborative Engineering Research Centers program.

Dr. Gabriel spent most of the year 1994 at the United States Senate Appropriations Committee, working as one of three majority professional staff members for the Subcommittee on VA, HUD, and Independent Agencies, chaired by Senator Barbara Mikulski. This subcommittee was responsible for appropriating about $90 billion annually among 25 federal organizations. Dr. Gabriel was also a researcher for six years at AT&T Bell Laboratories in New Jersey and spent six months in 1990 as a visiting professor at the University of Tokyo in Japan. She received her master's and doctorate degrees in electrical engineering and computer science from the Massachusetts Institute of Technology and her undergraduate electrical engineering degree from the University of Pittsburgh. She was an AT&T Bell Laboratories GRPW Fellow and a National Merit Scholar (Richard King Mellon Foundation). Her research publications focus on digital optical switching devices and systems exploiting ultra fast optical non-linearities in fibers and wave guides of glasses, polymers, and semiconductors, and she holds three patents.

LARRY KERR

Dr. Lawrence D. Kerr (Larry) is Director of Bioterrorism, Research and Development for the Office of Homeland Security (OHS) in the Executive Office of the President. Before joining OHS, he was a National Institutes of Health (NIH) agency representative to the National Science and Technology Council (NSTC) in the Office of Science and Technology Policy (OSTP). Dr. Kerr joined the Life Sciences division of OSTP in January 2001. He comes from his position as Chief of Transplantation, Transplantation and Immunology Branch at the National Institute of Allergy and Infectious Diseases (NIAID) at the NIH to serve as an advisor on science and technology.

Prior to his work at the NIH, Dr. Kerr worked in science and health care policy for Senator Orrin Hatch (R-UT) on the health subunit of the Senate Judiciary Committee during the 106th Congress. As a Robert Wood Johnson Fellow, he staffed the Senator on a variety of legislative affairs including: NIH reauthorization; medical device coding for Medicare reimbursement; radiation exposure compensation litigation; interagency coordination of counter-bioterrorism efforts; traumatic brain injury act, pediatric AIDS, and Ryan White CARE reauthorization.

In his capacity at OSTP, Dr. Kerr assisted the Director and other Executive Office of the President divisions in a variety of science and health care issues including: interagency coordination of chem/bio anti-terrorism technologies; infectious disease topics (HIV/AIDS, foot and mouth disease, etc.); the cloning of human beings; embryonic and adult stem cell biology; and administers the Presidential Early Career Awards for Scientists and Engineers (PECASE). His responsibilities include monitoring legislative activities along these subject areas.

As Director of Bioterrorism, R&D, Dr. Kerr assists the Senior Director for R&D and the Assistant to the President for Homeland Security in identifying and fostering policies to meet national objectives. Dr. Kerr represents OHS to other government agencies, participates in the development of short- and long-range policy alternatives, and interfaces with senior officials and staffs of the White House, Executive Office of the President, the Congress, the federal departments and agencies, and individuals from private industry and the academic community on antiterrorism programs that are responsive to the National Strategy on Homeland Security.

As an Assistant Professor in Microbiology and Immunology at Vanderbilt School of Medicine in Nashville, Tennessee, Dr. Kerr ran a basic science laboratory devoted to the study of the transcriptional regulation of gene products involved in HIV replication and breast cancer development. He has lectured at the national and international levels and received awards for teaching excellence. He is the author of more than 50 peer-reviewed articles, reviews, and book chapters. He holds a B.S. in Biology and Art History from the University of the South in Sewanee, TN. Dr. Kerr completed his Ph.D. in Cell Biology from Vanderbilt University in 1990 and undertook his post-doctoral work at the Salk Institute for Biological Studies in San Diego, California.

GORDON MOORE

Gordon E. Moore is currently Chairman Emeritus of Intel Corporation. Moore co-founded Intel in 1968, serving initially as Executive Vice President. He became President and Chief Executive Officer in 1975 and held that post until elected chairman and Chief Executive Officer in 1979. He remained CEO until 1987 and was named Chairman Emeritus in 1997.

Moore is widely known for "Moore's Law," in which he predicted that the number of transistors that the industry would be able to place on a computer chip would double every year. In 1995, he updated his prediction to once every two years. While originally intended as a rule of thumb in 1965, it has become the guiding principle for the industry to deliver ever-more-powerful semiconductor chips at proportionate decreases in cost.

Moore earned a B.S. in Chemistry from the University of California at Berkeley and a Ph.D. in Chemistry and Physics from the California Institute of Technology. He was born in San Francisco, California, on Jan. 3, 1929.

He is a director of Varian Associates, Gilead Sciences Inc., and Transamerica Corporation. He is a member of the National Academy of Engineering, a Fellow of the IEEE, and a Chairman of the Board of Trustees of the California Institute of Technology. He received the National Medal of Technology in 1990 from President George H. W. Bush.

SEAN O'KEEFE

Nominated by President George W. Bush and confirmed by the United States Senate, Sean O'Keefe was appointed by the President as the 10th Administrator of the National Aeronautics and Space Administration on December 21, 2001. As Administrator, O'Keefe leads the NASA team and manages its resources, as NASA seeks to advance exploration and discovery in aeronautics and space technologies.

O'Keefe joined the Bush Administration on inauguration day and served as the Deputy Director of the Office of Management and Budget until December 2001, overseeing the preparation, management, and administration of the Federal budget and government wide-management initiatives across the Executive Branch.

Prior to joining the Bush Administration, O'Keefe was the Louis A. Bantle Professor of Business and Government Policy, an endowed chair at the Syracuse University Maxwell School of Citizenship and Public Affairs. He also served as the Director of National Security Studies, a partnership of Syracuse University and Johns Hopkins University, for delivery of executive education programs for senior military and civilian Department of Defense managers. Appointed to these positions in 1996, he was previously Professor of Business Administration and Assistant to the Senior Vice President for Research and Dean of the Graduate School at the Pennsylvania State University.

Appointed as the Secretary of the Navy in July 1992 by President George H. W. Bush, O'Keefe previously served as Comptroller and Chief Financial Officer of the Department of Defense since 1989. Before joining Defense Secretary Dick Cheney's Pentagon management team in these capacities, he served on the United States Senate Committee on Appropriations staff for eight years, and was Staff Director of the Defense Appropriations Subcommittee. His public service began in 1978 upon selection as a Presidential Management Intern.

O'Keefe is a Fellow of the National Academy of Public Administration and has served as chair of an Academy panel on investigative practices. He was a Visiting Scholar at the Wolfson College of the University of Cambridge in the United Kingdom, a member of the Naval Postgraduate School's civil-military relations seminar team for emerging democracies and has conducted seminars for the Strategic Studies Group at Oxford University. He served on the national security panel to devise the 1988 Republican platform and was a member of the 1985 Kennedy School of Government program for national security executives at Harvard University.

In 1993, President Bush and Secretary Cheney presented him the Distinguished Public Service Award. He was also the recipient of the Department of the Navy's Public Service Award in December 2000. Sean O'Keefe was the 1999 faculty recipient of the Syracuse University Chancellor's Award for Public Service. He is the author of several journal articles, contributing author of *Keeping the Edge: Managing Defense for the Future*, released in October 2000, and in 1998, co-authored *The Defense Industry in the Post-Cold War Era: Corporate Strategies and Public Policy Perspectives*.

O'Keefe earned his Bachelor of Arts in 1977 from Loyola University in New Orleans, Louisiana, and his Master of Public Administration degree in 1978 from the Maxwell School. His wife Laura and children Lindsey, Jonathan, and Kevin, reside in northern Virginia.

RONALD M. SEGA

The Honorable Ronald M. Sega, Director of Defense Research and Engineering (DDR&E), is the chief technical advisor to the Secretary of Defense and the Under Secretary of Defense for Acquisition, Technology, and Logistics (USD-AT&L) for scientific and technical matters, basic and applied research, and advanced technology development. Dr. Sega also has management oversight for the Defense Advanced Research Projects Agency (DARPA).

Dr. Sega has had an extensive career in academia, research, and government service. He began his academic career as a faculty member in the Department of Physics at the U.S. Air Force Academy. His research activities in electromagnetic fields led to a Ph.D. in Electrical Engineering from the University of Colorado. He was appointed as Assistant Professor in the Department of Electrical and Computer Engineering at the University of Colorado at Colorado Springs in 1982. In addition to teaching and research activities, he also served as the Technical Director of the Laser and Aerospace Mechanics Directorate at the F. J. Seiler Research Laboratory and at the University of Houston as the Assistant Director of Flight Programs and Program Manager for the Wake Shield Facility. Dr. Sega became the Dean, College of Engineering and Applied Science, University of Colorado at Colorado Springs in 1996. Dr. Sega has authored or co-authored over 100 technical publications and was promoted to Professor in 1990. He is also a Fellow of the Institute of Electrical and Electronic Engineers and the Institute for the Advancement of Engineering.

In 1990, Dr. Sega joined NASA, becoming an astronaut in July 1991. He served as a mission specialist on two Space Shuttle Flights, STS-60 in 1994, the first joint U.S. Russian Space Shuttle Mission and the first flight of the Wake Shield Facility, and STS-76 in 1996, the third docking mission to the Russian space station Mir where he was the Payload Commander. He was also the Co-Principal Investigator for the Wake Shield Facility and the Director of Operations for NASA activities at the Gagarin Cosmonaut Training Center, Russia, from 1994 to 1995.

Dr. Sega has also been active in the Air Force Reserves. A Command Pilot in the Air Force with over 4,000 hours, he has served in various operational flying assignments, including a tour of duty as an Instructor Pilot. From 1984 to 2001, as a reservist assigned to Air Force Space Command (AFSPC), he held positions in planning analysis and operational activities, including Mission Ready Crew Commander for satellite operations—Global Positioning System (GPS)—Defense Support Program (DSP), and Midcourse Space Experiment (MSX), etc. He was promoted to the rank of Major General in the Air Force Reserves in July 2001.

WILLIAM SPENCER

Currently the Chairman of the Board of SEMATECH, Dr. Spencer served as the President and Chief Executive Officer of the consortium from 1990 through 1996. SEMATECH is a research and development consortium based in Austin, Texas jointly funded by the semiconductor industry member companies and the U.S. government. It was established in 1987 to solve the technical challenges required to maintain U.S. leadership in the global semiconductor industry. Before joining SEMATECH, Dr. Spencer was group vice president and senior technical officer at Xerox Corporation in Stamford, Conn. He has also served as Vice President of Xerox Palo Alto Research Center, Director of Systems Development at Sandia National Laboratories in Livermore, and Director of Microelectronics at Sandia National Laboratories in Albuquerque. He began his career at Bell Telephone Laboratories.

Dr. Spencer received an A.B. degree from William Jewell College in Liberty, Missouri, followed by an M.S. degree in Mathematics and a Ph.D. in Physics from Kansas State University. He was awarded the Regents Meritorious Service Medal from the University of New Mexico in 1981, and an honorary doctorate degree from William Jewell College in 1990. He is a member of the National Academy of Engineering, a Fellow of IEEE, and serves on numerous advisory groups and boards, including the Board on Science, Technology, and Economic Policy.

JAMES TURNER

James Turner has served on the professional staff of the Committee on Science in the U.S. House of Representatives for approximately 20 years. He currently serves as the Full Committee Chief Democratic Counsel where he works across the board on the Committee's legislative agenda.

For the 10 years prior to the Republican takeover of Congress, Mr. Turner was the Committee's senior staff member for technology policy including four years as technology subcommittee staff director. He also served as a subcommittee legal counsel. During the late 1970s and early 1980s, Mr. Turner worked on the Committee's Republican staff as Minority Energy Counsel.

During his years on the Committee, Mr. Turner has worked on numerous bills, reports, and hearings on a wide variety of topics. These include the international competitiveness of U.S. industry, environmental and energy research and development, trade and technology policy, intellectual property, standards, and technology transfer.

Mr. Turner also spent 3 years working for Wheelabrator-Frye, 2 years for Congressman Gary Myers, 2 years for the State of Connecticut, and shorter periods with NASA and FAA. He holds degrees from Georgetown and Yale Universities and from Westminster College and attended the Senior Managers in Government Program at Harvard.

PATRICK WINDHAM

Until April 1997, Patrick Windham served as Senior Professional Staff Member for the Subcommittee on Science, Technology, and Space of the U.S. Senate's Committee on Commerce, Science, and Transportation. He helped the Senators oversee and draft legislation for several major civilian R&D agencies with responsibility for science, technology, and U.S. competitiveness; industry-government-university R&D partnerships; state economic development; federal laboratory technology transfer; high-performance computing; and computer encryption. From 1982 to 1984, he served as a legislative aide in the personal office of Senator Ernest Hollings. From 1976 to 1978, he worked as a Congressional fellow with the Senate Commerce Committee, and then returned to California from 1978 to 1982 to pursue further graduate studies in political science at the University of California at Berkeley.

Mr. Windham holds a Master's of Public Policy from the University of California at Berkeley and a B.A. from Stanford University. He is currently an independent, California-based consultant on science and technology policy issues.

Appendix B

Participants List*

Fred Adler
WDC USA World-Wide

Bruce Alberts
National Academy of Sciences

Jeff Alexander
Washington-CORE

Anne K. Altman
IBM Corporation

William Anderson
U.S. Government

Kathy Behrens
RS Investment Management

Arden Bement
National Institute of Standards and Technology

Tabitha Benney
The National Academies

Richard Bissell
The National Academies

Raymond Blair
IBM Corporation

Sherwood L. Boehlert
U.S. House of Representatives

William Boger
Perkins, Smith, Cohen & Crowe, LLP

William B. Bonvillian
Office of Senator Lieberman

Michael Borrus
The Petkevich Group, LLC

---

*Speakers in italics.

APPENDIX B

Matthew Burrows
U.S. Government

Jennifer Buxe
IDA

Joanne P. Carney
American Association for the
   Advancement of Science

Robert Carpenter
University of Maryland, Baltimore
   County

Gail Cassell
Lilly Research Laboratories
Eli Lilly & Company

Mark Coburn
University of Rochester

E. William Colglazier
The National Academies

Camille Collett
The National Academies

Marc Collett
ViroPharma Incorporated

Karen Conti
Epsilon Systems Solutions

Brian Costello
Edmond Scientific Company

Terence Costello
The Eagle Group

Chris Daly
IBM Corporation

Alessandro Damiani
Delegation of the European
   Commission

Warren DeVries
National Science Foundation

David Dierksheide
The National Academies

Sidney Draggan
Environmental Protection Agency

Michael Eichberg
American Chemical Society

Kerstin Eliasson
Embassy of Sweden

Gerald Etzold
National Security Agency

Lauren Ewald
IBM Corporation

Tara Federici
Advanced Medical Technology
   Association

Maryann Feldman
Johns Hopkins University

Kenneth Flamm
University of Texas at Austin

Steve Flynn
Council on Foreign Relations

---

*Speakers in italics.

Nancy Forbes
U.S. Government

Forrest Frank
Institute for Defense Analyses

Peter Freeman
National Science Foundation

Cita Furlani
Advanced Technology Program
National Institute of Standards and
 Technology

Christina Gabriel
Carnegie Mellon University

William Gaines
Office of Science and Technology
 Policy

John Gardenier
Centers for Disease Control

Turkan Gardenier
Pragmatica Corp.

Ricky Garris
IBM Corporation

Gradimir Georgevich
Advanced Technology Program
National Institute of Standards and
 Technology

Jeffrey Goldman
American Institute of Biological
 Sciences

Donald Goldstein
Institute for Defense Analyses

David Goldston
House Science Committee

Dan Greeburg
The Lancet

Tim Hackman
IBM Governmental Programs

Gerald Hane
Globalvation

Mark Harkins
House Science Committee

Phillip Harman
Pennsylvania Avenue Associates

Chris Hayter
The National Academies

Carole Heilman
National Institute of Allergy and
 Infectious Diseases

Robert Hershey
Robert L. Hershey, P.E.

John B. Horrigan
Pew Internet Project

Kent Hughes
The Woodrow Wilson Center

Kevin Hurst
Office of Science and Technology
 Policy

---

*Speakers in italics.

APPENDIX B

Richard Johnson
Arnold & Porter

Steve Johnson
Cornell University

Teresa Jones
NACFAM

Dale Jorgenson
Harvard University

Jim Kadtke
Office of Senator John Warner

Nina Kaull
The National Academies

Dmitriy Kazakov
Russian Trade Representation

Larry Kerr
Department of Homeland Security

B. Lee Kindberg
Institute for Defense Analyses

Kathleen Kingscott
IBM Corporation

Aaron Kirtley
Advanced Technology Program
National Institute of Standards and Technology

Karl Koehler
National Science Foundation

Adam Korobow
The National Academies

David Kramer
Science and Government Report

Richard Lambert
Department of Health and Human Services

Thomas Libert
The National Academies

Mark Lippman
IBM Global Services

William Long
Business Performance Research Associates

Clark McFadden
Dewey Ballantine

Al Mecchi

Arthur Melmed
George Mason University

Stephen Merrill
The National Academies

C. Bradley Moore
Ohio State University

Gordon Moore
Intel Corporation

Evan Morris
Patton Boggs

Michael Mowatt

---

*Speakers in italics.

Russel Moy
The National Academies

Scott Nance
New Technology Week

Norman Neureiter
Department of State

William New
National Journal

Arnauld Nicogossian
National Aeronautics and Space
  Administration

Jeannette Pohl Nielsen
Embassy of Denmark

Markku Oikarainen
TEKES

Sean O'Keefe
National Aeronautics and Space
  Administration

Omid Omidvar
Advanced Technology Program
National Institute of Standards and
  Technology

Kaare Pedersen
Embassy of Denmark

David Y. Peyton
National Association of
  Manufacturers

John Pitale
Edmond Scientific Company

Jeanne Powell
Advanced Technology Program
National Institute of Standards and
  Technology

Darin Powers
RAE, Inc.

Katie Rahmlow
House Science Committee

Lisette Ramcharan
Embassy of Canada

Samuel M. Rankin III
American Mathematical Society

Alan Rapoport
National Science Foundation

Andrew Reynolds
Department of State

Steven Rizzi
SAIC

Claire Saundry
National Institute of Standards and
  Technology

Craig Schultz
The National Academies

Ronald Sega
Defense of Defense

Arun Seraphin
Office of Senator Lieberman

Sujai Shivakumar
The National Academies

---

*Speakers in italics.

Elissa Sobolewski
Advanced Technology Program
National Institute of Standards and
    Technology

Anne Solomon
Center for Strategic and International
    Studies

William Spencer
International SEMATECH

Richard Spivack
Advanced Technology Program
National Institute of Standards and
    Technology

Marc Stanley
Advanced Technology Program
National Institute of Standards and
    Technology

Todd Stewart
Ohio State University

Ronald Stoltz
Sandia National Laboratories

Istvan Takacs
Embassy of the Republic of Hungary

Kevin Thomas
George Mason University

Jean Toal-Eisen
Office of Senator Hollings

James Turner
House Science Committee

Richard Van Atta
Institute for Defense Analyses

Frances Velez
KPMG Consulting

Steven Wallach
Pennie & Edmonds LLP

Charles Wessner
The National Academies

Patrick Windham
Windham Consulting

Jack Yadvish
National Aeronautics and Space
    Administration

---

*Speakers in italics.

Appendix C

Bibliography

Acs, Zoltan J., and David B. Audretsch. 1991. *Innovation and Small Firms.* Cambridge: MIT Press.

Adams, Chris. 2001. "Laboratory Hybrids: How Adroit Scientists Aid Biotech Companies with Taxpayer Money—NIH Grants Go to Non-profits Tied to For-profit Firms Set up by Researchers," *Wall Street Journal.* New York: Dow Jones and Company. January 30. p. A1.

Aizcorbe, Ana Kenneth Flamm, and Anjum Khurshid. 2002. "The Role of Semiconductor Inputs in IT Hardware Price Decline: Computers vs. Communications." Washington, D.C.: Federal Reserve Board. Finance and Economics Discussion Series. August.

Ambrose, Stephen. 2000. *Nothing Like It in the World: The Men Who Built the Transcontinental Railroad 1863–1869.* New York: Simon and Schuster.

American Association for the Advancement of Science. 2002. *AAAS Preliminary Analysis of R&D in FY 2003 Budget.* February 8. <www.aaas.org/spp/R&D>.

Audretsch, David B. 1995. *Innovation and Industry Evolution.* Cambridge: The MIT Press.

Audretsch, David B., and Roy Thurik. 1999. *Innovation, Industry, Evolution, and Employment.* Cambridge: Cambridge University Press.

Audretsch, David B., Barry Bozeman, Kathryn L. Combs, Maryann Feldman, Albert N. Link, Donald S. Siegel, Paula Stephan, Gregory Tassey, and Charles Wessner. 2002. "The Economics of Science and Technology." *Journal of Technology Transfer* 27:155–203.

Baily, M. N., and A. Chakrabati. 1998. *Innovation and the Productivity Crisis.* Washington, D.C.: The Brookings Institution.

Baily, M. N., and R. Z. Lawrence. 2001. "Do We Have an E-conomy?" NBER Working Paper 8243. April 23.

Bilstein, Roger E. 1989. *A History of the NACA and NASA, 1915–1990.* Washington, D.C.: National Aeronautics and Space Administration.

Bingham, Richard. 1998. *Industrial Policy American Style: From Hamilton to HDTV.* New York: M. E. Sharpe.

Borrus, Michael. 1997. Testimony before the U.S. House of Representatives Committee on Science: Subcommittee on Technology. April 10.

Borrus, Michael, and Jay Stowsky. 1997. "Technology Policy and Economic Growth." BRIE Working Paper 97. April.

Brander, J. A., and B. J. Spencer. 1983. "International R&D Rivalry and Industrial Strategy." *Review of Economic Studies* 50:707–722.

Brander, J. A., and B. J. Spencer, 1985. "Export Subsidies and International Market Share Rivalry." *Journal of International Economics* 16:83–100.

Branscomb, L. 2001. Testimony before U.S. House of Representatives Committee on Science: Subcommittee on Technology. June 14.

Branscomb, L., and P. Auerswald. 2001. *Taking Technical Risk: How Innovators, Executives, and Investors Manage High-Tech Risks.* Cambridge: The MIT Press.

Branscomb, L. M., and J. Keller, eds. 1998. *Investing in Innovation: Creating a Research and Innovation Policy.* Cambridge: The MIT Press.

Brown, George, and James Turner. 1999. "The Federal Role in Small Business Research." *Issues in Science and Technology* 15(4):51–58.

Brown, Martin. 1995. *Impacts of National Technology Programs.* Paris: Organization for Economic Cooperation and Development.

Browning, L. D., J. M. Beyer, and J. C. Shetler. 1995. "Building cooperation in a competitive industry: SEMATECH and the semiconductor industry." *Academy of Management Journal* 38(1):113–151.

Browning, Larry D., and Judy C. Shetler. 2000. *SEMATECH: Saving the U.S. Semiconductor Industry.* College Station: Texas A&M University Press.

Cahners In-Stat Group. 1999. "Is China's Semiconductor Industry Market Worth the Risk for Multinationals? Definitely!" *Cahners In-Stat Group* March 29.

Campbell, Donald E. 1995. *Incentives: Motivations and the Economics of Information.* Cambridge: Cambridge University Press.

Chandler, Alfred P. 1962. *Strategy and Structure: Chapters in History of the Industrial Enterprise.* Cambridge: The MIT Press.

Chesbrough, Hank. 2001. "Is the Central R&D Lab Obsolete?" *Technology Review* April 24.

Coburn, Christopher, and Dan Berglund. 1995. *Partnerships: A Compendium of State and Federal Cooperative Technology Programs.* Columbus: Battelle Press.

Cohen, Linda R., and Roger G. Noll. 1991. *The Technology Pork Barrel.* Washington, D.C.: The Brookings Institution.

Council of Economic Advisers. 1995. *Economic Report of the President.* Washington, D.C.: U.S. Government Printing Office. January.

Council of Economic Advisers. 1995. *Supporting Research and Development to Promote Economic Growth: The Federal Government's Role.* Washington, D.C.: U.S. Government Printing Office.

Council of Economic Advisers. 2000. *The Annual Report of the Council of Economic Advisers.* Washington D.C.: U.S. Government Printing Office.

Council of Economic Advisers. 2001. *Economic Report of the President.* Washington, D.C.: U.S. Government Printing Office. January.

Council of Economic Advisers. 2002. *Economic Report of the President.* Washington, D.C.: U.S. Government Printing Office. January.

David, Paul. 2000. "Understanding Digital Technology's Evolution and the Path of Measured Productivity Growth: Present and Future in the Mirror of the Past." In E. Brynjolfsson and Brian Kahin, eds. *Understanding the Digital Economy: Data, Tools, and Research.* Cambridge: The MIT Press.

David, Paul A., Bronwyn H. Hall, and Andrew A. Toole. 1999. "Is Public R&D a Complement or Substitute for Private R&D? A Review of the Econometric Evidence." NBER Working Paper 7373, October.

Davis, Steven J., John Haltiwanger, and Scott Schuh. 1994. "Small Business and Job Creation: Dissecting the Myth and Reassessing the Facts." *Business Economics* 29(3):113–122.

de Tocqueville, Alexis. 2000. *Democracy in America.* Chicago: University of Chicago Press.
Diebold Jr., William. 1980. "Past and Future Industrial Policy in the United States." In J. Pinder, ed., *National Industrial Strategies and the World Economy.* London: Allanheld, Osmun & Company.
The Economist. 1989. "The Rise and Fall of America's Small Firms." *The Economist* January 21.
The Economist. 2000. "A Thinker's Guide." *The Economist* March 30.
The Economist. 2001. "The Great Chip Glut." *The Economist* August 11.
The Economist. 2001. "Protein Based Computer Memories, Data Harvest." *The Economist* December 22.
Evanson, Robert E., and Wallace E. Huffman. 1993. *Science for Agriculture: A Long-term Perspective.* Ames: Iowa State University Press.
Fallows, J. 1994. *Looking into the Sun: The Rise of the New East Asian Economic and Political System.* New York: Pantheon Books
Flamm, Kenneth. 1988. *Creating the Computer.* Washington, D.C.: The Brookings Institution.
Flamm, Kenneth. 1996. *Mismanaged Trade? Strategic Policy and the Semiconductor Industry.* Washington, D.C.: The Brookings Institution.
Flax, Alexander. 1999. National Academy of Engineering. Personal Communication. September.
Fogel, Robert W. 1964. *Railroads and American Economic Growth: Essays in Econometric History.* Baltimore: Johns Hopkins University Press.
Galbraith, John Kenneth. 1957. *The New Industrial State.* Boston: Houghton Mifflin.
Graham, Otis L. 1992. *Losing Time: The Industrial Policy Debate.* Cambridge: Harvard University Press.
Greenspan, Alan. 2000. Remarks before the *White House Conference on the New Economy.* Washington, D.C. April 5.
Griliches, Zvi. 1990. *The Search for R&D Spillovers.* Cambridge: Harvard University Press.
Grindley, Peter, David C. Mowery, and Brian Silverman. 1994. "SEMATECH and Collaborative Research: Lessons in the Design of High-Technology Consortia." *Journal of Policy Analysis and Management* 13(4):723–58.
Grossman, Gene, and Elhanan Helpman. 1993. *Innovation and Growth in the Global Economy.* Cambridge: The MIT Press.
Hart, David M. 1998. *Forged Consensus: Science, Technology, and Economic Policy in the United States, 1921–1953.* Princeton: Princeton University Press.
Horrigan, John B. 1999. "Cooperating Competitors: A Comparison of MCC and SEMATECH." Monograph. Washington, D.C.: National Research Council.
Hounshell, David A. 1985. *From the American System to Mass Production, 1800–1932.* Baltimore: Johns Hopkins University Press.
Hudgins, Edward L. 1995. Testimony before the Senate Committee on Commerce, Science, and Transportation. August 1.
International SEMATECH. 2002. *Annual Report 2001.* Austin: International SEMATECH.
Jarboe, Kenan Patrick, and Robert D. Atkinson. 1998. *The Case for Technology in the Knowledge Economy; R&D, Economic Growth and the Role of Government.* Washington, D.C.: Progressive Policy Institute. June 1.
Johnson, Chalmers. 1982. *MITI and the Japanese Miracle: The Growth of Industrial Policy 1925–1975.* Stanford: Stanford University Press.
Jorgensen, Dale W. 2001. "Information Technology and the U.S. Economy." Presidential Address to the American Economic Association. New Orleans, LA. January 6.
Jorgenson, Dale, and Kevin Stiroh. 2000. "Raising the Speed Limit: U.S. Economic Growth in the Information Age." *Brookings Papers-on-Economic-Activity.* Washington, D.C.: The Brookings Institution. pp. 125–211.
Kelly, Henry, et al. 2002. "Training Technology Against Terror: Using Advanced Technology to Prepare America's Emergency Medical Personnel and First Responders for a Weapon of Mass Destruction Attack." Washington, D.C.: Federation of American Scientists. September.

Kenney, Martin, ed. 2000. *Understanding Silicon Valley: The Anatomy of an Entrepreneurial Region.* Stanford: Stanford University Press.

Kleinman, Daniel Lee. 1995. *Politics on the Endless Frontier: Postwar Research Policy in the United States.* Durham: Duke University Press.

Koizumi, Kei, and Paul W. Turner. 2002. *Congressional Action on Research and Development in the FY 2002 Budget.* Washington, D.C.: American Association for the Advancement of Science.

Kornai, Janos. 1980. *Economics of Shortage.* Amsterdam: North Holland.

Krugman, P. Undated. "Some Chaotic Thoughts on Regional Dynamics." at <http://www.wws.princeton.edu/~pkrugman/temin.html>.

Krugman, P. 1990. *Rethinking International Trade.* Cambridge: MIT Press.

Krugman, P. 1991. *Geography and Trade.* Cambridge: MIT Press.

Krugman, P. 1994. *Peddling Prosperity: Economic Sense and Nonsense in an Age of Diminished Expectations.* New York: W.W. Norton Press.

Langlois, Richard N. 1991. "Schumpeter and the Obsolescence of the Entrepreneur." Working Paper, 91-1503. University of Connecticut Department of Economics. November.

Langlois, Richard N., and Paul L. Robertson. 1996. "Stop Crying over Spilt Knowledge: A Critical Look at the Theory of Spillovers and Technical Change." Paper prepared for the MERIT Conference on Innovation, Evolution, and Technology. Maastricht, Netherlands. August 25–27.

Larson, Charles F. 2000. "The Boom in Industry Research." *Issues in Science and Technology* 16(4):27.

Lebow, Irwin. 1995. *Information Highways and Byways.* New York: Institute of Electrical and Electronics Engineers.

Linden, Greg, David Mowery, and Rosemarie Ziedonis. 2001. "National Technology Policy in Global Markets." In Albert Link and Maryann Feldman, eds. *Innovation Policy in the Knowledge-based Economy.* Boston: Kluwer Academic Publishers.

Link, A. N. 1996. "Research Joint Ventures: Patterns From Federal Register Filings." *Review of Industrial Organization.* 11(5):617–628.

Link, A. N. 1999. "Public/Private Partnerships as a Tool to Support Industrial R&D: Experiences in the United States." *Final Report to the Working Group on Innovation and Technology Policy of the OECD Committee for Scientific and Technology Policy.* January.

Luger, Michael I., and Harvey A. Goldstein. 1991. *Technology in the Garden; Research Parks & Regional Economic Development.* Chapel Hill: University of North Carolina Press.

Mann, Catherine. 1993. *Is the U.S. Trade Deficit Sustainable?* Washington, D.C.: Institute for International Economics.

Mann, Charles C. 2000. "The End of Moore's Law?" *Technology Review* May/June.

Mansfield, Edwin. 1985. "How Fast Does New Industrial Technology Leak Out?" *Journal of Industrial Economics* 34(2):217–224.

Mansfield, Edwin. 1991. "Academic Research and Industrial Innovation" *Research Policy* February.

Marshall, Alfred. 1920. *Industry and Trade.* 3rd edition. London: Macmillan.

Martin, Brookes, and Zaki Wahhaj. 2000. "The Shocking Economic Impact of B2B." *Global Economic Paper.* 37. Goldman Sachs. February 3.

May, John. 2002. "Angel Alliances and Angel Practices." Presented at the State of the Angel Market Workshop. Boston, MA. March 27.

McCraw, Thomas. 1986. "Mercantilism and the Market: Antecedents of American Industrial Policy." In *The Politics of Industrial Policy.* Claude E. Barfield and William A. Schambra, eds. Washington, D.C.: American Enterprise Institute for Public Policy Research.

McKinsey Global Institute. 2001. *U.S. Productivity Growth 1995–2000, Understanding the Contribution of Information Technology Relative to Other Factors.* Washington, D.C.: McKinsey & Company. October.

Merrill, Stephen A., and Michael McGeary. 1999. "Who's Balancing the Federal Research Portfolio and How?" *Science* 285(September 10):1679–1680.

Middendorf, William H. 1981. *What Every Engineer Should Know About Inventing.* New York and Basel: Marcel Dekker Inc.

Moore, Gordon E. 1965. "Cramming More Components onto Integrated Circuits." *Electronics* 38(8, April):19.

Moore, Gordon E. 1997. "The Continuing Silicon Technology Evolution Inside the PC Platform." *Intel Developer Update* Issue 2. October 15.

Mowery, David. 1998. "Collaborative R&D: How Effective Is It?" *Issues in Science and Technology* 15(1):37–44.

Mowery, David, and N. Hatch. 2002. "Managing the Development and Introduction of New Manufacturing Processes in the Global Semiconductor Industry." In G. Dosi, R. Nelson, and S. Winter, eds. *The Nature and Dynamics of Organizational Capabilities.* New York: Oxford University Press.

Mowery, David, and Nathan Rosenberg. 1989. *Technology and the Pursuit of Economic Growth.* Cambridge: Cambridge University Press.

Mowery, David, and Nathan Rosenberg. 1998. *Paths of Innovation: Technological Change in 20th Century America.* New York: Cambridge University Press.

Mowery, David, and Brian Silverman. 1996. "SEMATECH and Collaborative Research: Lessons in the Design of High-Technology Consortia." *Journal of Policy Analysis and Management* 13(4).

Nadiri, Ishaq. 1993. *Innovations and Technological Spillovers.* NBER Working Paper No. 4423.

Nance, Scott. 2000. "Broad Federal Research Required to Keep Semiconductors on Track." *New Technology Week* October 30.

National Bureau of Standards. 1977. *The Influence of Defense Procurement and Sponsorship of Research and Development on the Development of the Civilian Electronics Industry.* National Bureau of Standards. June 30.

National Research Council. 1992. *The Government Role in Civilian Technology: Building A New Alliance.* Washington, D.C.: National Academy Press.

National Research Council. 1996. *Conflict and Cooperation in National Competition for High Technology Industry.* Washington, D.C.: National Academy Press.

National Research Council. 1999. *The Advanced Technology Program: Challenges and Opportunities.* Charles W. Wessner, ed. Washington, D.C.: National Academy Press.

National Research Council. 1999. *Funding a Revolution; Government Support for Computing Research.* Washington, D.C.: National Academy Press.

National Research Council. 1999. *Industry-Laboratory Partnerships: A Review of the Sandia Science and Technology Park Initiative.* Charles W. Wessner, ed. Washington, D.C.: National Academy Press.

National Research Council. 1999. *New Vistas in Transatlantic Science and Technology Cooperation.* Charles W. Wessner, ed. Washington, D.C.: National Academy Press.

National Research Council. 1999. *The Small Business Innovation Research Program: Challenges and Opportunities.* Charles W. Wessner, ed. Washington, D.C.: National Academy Press.

National Research Council. 1999. *U.S. Industry in 2000: Studies in Competitive Performance.* Washington, D.C.: National Academy Press.

National Research Council. 2000. *The Small Business Innovation Research Program: An Assessment of the Department of Defense Fast Track Initiative.* Charles W. Wessner, ed. Washington D.C.: National Academy Press.

National Research Council. 2001. *The Advanced Technology Program: Assessing Outcomes.* Charles W. Wessner, ed. Washington D.C.: National Academy Press.

National Research Council. 2001. *A Review of the New Initiatives at the NASA Ames Research Center.* Charles W. Wessner, ed. Washington D.C.: National Academy Press.

National Research Council. 2001. *Review of the Research Program of the Partnership for a New Generation of Vehicles: Seventh Report.* Washington, D.C.: National Academy Press.

National Research Council. 2001. *Trends in Federal Support of Research and Graduate Education.* Stephen A. Merrill, ed. Washington, D.C.: National Academy Press.
National Research Council. 2002. *Capitalizing on New Needs and New Opportunities: Government-Industry Partnerships in Biotechnology and Information Technologies.* Charles W. Wessner, ed. Washington D.C.: National Academy Press.
National Research Council. 2002. *Making the Nation Safer: The Role of Science and Technology in Countering Terrorism.* Washington D.C.: The National Academies Press.
National Research Council. 2002. *Measuring and Sustaining the New Economy.* D. Jorgenson and C. Wessner, eds. Washington, D.C.: National Academy Press.
National Research Council. 2002. *Partnerships for Solid-State Lighting.* Charles W. Wessner, ed. Washington D.C.: National Academy Press.
National Research Council. 2002. *Small Wonders, Endless Frontiers: A Review of the National Nanotechnology Initiative.* Washington, D.C.: National Academy Press.
National Research Council. 2003. *Securing the Future: Regional and National Programs to Support the Semiconductor Industry.* Charles W. Wessner, ed. Washington D.C.: The National Academies Press.
National Research Council. 2003. *Government-Industry Partnerships for the Development of New Technologies: Summary Report,.* C. Wessner, ed., Washington D.C.: The National Academies Press.
National Science Board. 1998. *Science and Engineering Indicators, 1998.* Arlington: National Science Foundation.
Nelson, Richard R. 1982. *Government and Technological Progress.* New York: Pergamon Press.
Nelson, Richard R. 1993. *National Innovation Systems.* New York: Oxford University Press.
Nelson, Richard R. 2000. *The Sources of Economic Growth.* Cambridge: Harvard University Press.
Nikkei Microdevices. 2001. "From Stagnation to Growth, The Push to Strengthen Design." *Nikkei Microdevices* January.
Nikkei Microdevices. 2001. "Three Major European LSI Makers Show Stable Growth Through Large Investments." *Nikkei Microdevices* January.
Noll, Roger. 2002. "Federal R&D in the Anti-Terrorist Era." In *Innovation Policy and the Economy*, Vol. 3. Adam B. Jaffe, Joshua Lerner and Scott Stern, eds. Cambridge: The MIT Press.
North, Douglass C. 1990. *Institutions, Institutional Change, and Economic Performance.* Cambridge: Cambridge University Press.
Office of Science and Technology Policy. 2000. *Fact Sheet on How Federal R&D Investments Drive the U.S. Economy.* Washington, D.C.: Executive Office of the President. June 15.
Okimoto, Daniel I. 1989. *Between MITI and the Market: Japanese Industrial Policy for High Technology.* Stanford: Stanford University Press.
Oliner, Stephen, and Daniel Sichel. 2000. "The Resurgence of Growth in the Late 1990s: Is Information Technology the Story?" *Journal of Economic Perspectives* 14(4, Fall).
Olson, Mancur. 1965. *The Logic of Collective Action.* Cambridge: Harvard University Press.
Organization for Economic Cooperation and Development. 1997. *Small Business Job Creation and Growth: Facts, Obstacles, and Best Practices.* Paris: Organization for Economic Cooperation and Development.
Organization for Economic Cooperation and Development. 2000. *Is There a New Economy? A First Report on the OECD Growth Project.* Paris: Organization for Economic Cooperation and Development.
Ostrom, Elinor. 1998. "A Behavioral Approach to the Rational Choice Theory of Collective Action." *American Political Science Review* 92(1).
Ostrom, Vincent. 1991. *The Meaning of American Federalism: Constituting a Self-Governing Society.* San Francisco: Institute for Contemporary Studies Press.
Packan, Paul A. 1999. "Pushing the Limits: Integrated Circuits Run into Limits Due to Transistors." *Science* September 24.

Polanyi, Michael. 1951. *Logic of Liberty*. Chicago: University of Chicago Press.
Porter, Michael. 1998. "Clusters and the New Economics of Competition." *Harvard Business Review* November–December.
Porter, Michael. 2001. *Clusters of Innovation: Regional Foundations of Competitiveness.* Washington, D.C.: Council on Competitiveness.
Prestowitz, Clyde. 1988. *Trading Places*. New York: Basic Books.
Procassini, Andrew A. 1995. *Competitors in Alliance: Industrial Associations, Global Rivalries, and Business-Government Relations.* Westport: Quorum Books.
Rodgers, T.J. 1998. "Silicon Valley Versus Corporate Welfare." *CATO Institute Briefing Papers.* Briefing Paper No. 37. 27 April.
Romer, Paul. 1990. "Endogenous Technological Change." *Journal of Political Economy* 98(5):71–102.
Rosenberg, Nathan. 1982. *Inside the Black Box: Technology and Economics.* New York: Cambridge University Press.
Rosenbloom, Richard, and William Spencer. 1996. *Engines of Innovation: U.S. Industrial Research at the End of an Era.* Boston: Harvard Business School Press.
Ruttan, Vernon W. 2001. *Technology, Growth and Development: An Induced Innovation Perspective.* New York: Cambridge University Press.
Samuel, Richard. 1994. *Rich Nation, Strong Army: National Security and Technological Transformation of Japan.* Ithaca: Cornell University Press.
Saxenian, AnnaLee. 1994. *Regional Advantage: Culture and Competition in Silicon Valley and Route 128.* Cambridge: Harvard University Press.
Schaller, Robert R. 1999. "Technology Roadmaps: Implications for Innovation, Strategy, and Policy." Ph.D. Dissertation Proposal, Institute for Public Policy, George Mason University.
Scherer, F. M. 1999. *New Perspectives on Economic Growth and Technological Innovation.* Washington D.C.: Brookings.
Schumpeter, Joseph A. 1950. *Capitalism, Socialism and Democracy.* New York: Harper and Row.
Semiconductor Industry Association. 2001. *2001 Annual Databook: Review of Global and U.S. Semiconductor Competitive Trends, 1978–2000.* San Jose: Semiconductor Industry Association.
Silber & Associates. 1996. *Survey of Advanced Technology Program 1990–1992 Awardees: Company Opinion about the ATP and Its Early Effects.* NIST GCR 97-707. February.
Smith, Hedrick. 1995. *Rethinking America.* New York: Random House.
Sohl, Jeffrey E. 1999. "The Early-Stage Equity Market in the USA." *Venture Capital* 1(2):101–120.
Sohl, Jeffery E. 2002. "The Private Equity Market in the U.S.: What a Long Strange Trip It Has Been." Mimeo. University of New Hampshire: Whittemore School of Business and Economics.
Solomon Associates. 1993. *Advanced Technology Program: An Assessment of Short-Term Impacts—First Competition Participants.* Solomon Associates. February.
Solow, Robert S. 1957. "Technical Change and the Aggregate Production Function." *Review of Economics and Statistics* 39:312–320.
Sporck, Charles E. (with Richard L. Molay). 2001. *Spinoff: A Personal History of the Industry That Changed the World.* Sarnac Lake: Sarnac Lake Publishing.
Stowsky, Jay. 1996. "Politics and Policy: The Technology Reinvestment Program and the Dilemmas of Dual Use." Mimeo. University of California.
Strategic Marketing Associates. 2002. *World Fab Watch.* Santa Cruz: World Fab Watch.
Susaki, Hajime. 2000. "Japanese Semiconductor Industry's Competitiveness: LSI Industry in Jeopardy." *Nikkei Microdevices* December.
Tennenhouse, David. 2002. Joint Strategic Assessments Group and Defense Advanced Research Projects Agency conference, *The Global Computer Industry Beyond Moore's Law: A Technical, Economic, and National Security Perspective.* Herndon, VA. January 14–15.
Thompson, Robert Luther. 1947. *Wiring a Continent: The History of the Telegraph Industry in the United States 1823–1836.* Princeton: Princeton University Press.

APPENDIX C

Trauthwein, Christina. 2001. "You Say You Want a Revolution . . . ." *Architectural Lighting* May.

Tyson, Laura. 1992. *Who's Bashing Whom? Trade Conflict in High Tech Industries.* Washington, D.C.: Institute for International Economics.

U.S. Commission on National Security/21st Century. 2001. *Roadmap for National Security: Imperative for Change.* January.

U.S. Department of Commerce. 2002. *The Advanced Technology Program: Reform with a Purpose.* Washington, D.C.: U.S. Department of Commerce. February.

U.S. Department of Energy. 1995. *Alternative Futures for the Department of Energy National Laboratories.* The "Galvin Report." Washington, D.C.: U.S. Department of Energy.

U.S. General Accounting Office. 1992. *SEMATECH's Technological Progress and Proposed R&D Program.* GAO/RCEED/92-223 BR. Washington D.C.: U.S. General Accounting Office. July.

U.S. Senate Committee on Appropriations. 1998. Report from the Committee on Appropriations to accompany Bill S. 2260. Washington, D.C.: United States Senate.

U.S. Senate Committee on Appropriations. 1998. *Senate Report 105-235.* Departments of Commerce, Justice, and State, the Judiciary, and Related Agencies Appropriation Bill. Washington, D.C.: United States Senate.

Varmus, Harold. 1999. "The Impact of Physics on Biology and Medicine." Plenary Talk, Centennial Meeting of the American Physical Society. Atlanta. March 22.

Vonortas, N. S. 1997. Cooperation in Research and Development. Norwell: Kluwer Academic Publishers.

Wolff, Alan Wm., Thomas R. Howell, Brent L. Bartlett, and R. Michael Gadbaw. eds. 1995. *Conflict Among Nations: Trade Policies in the 1990s.* San Francisco: Westview Press.

The World Bank. 1993. *The East Asian Economic Miracle: Economic Growth and Public Policy.* Policy Research Report. New York: Oxford University Press.

Zachary, G. Paschal. 1997. *Endless Frontier: Vannevar Bush, Engineer of the American Century.* New York: The Free Press.

Zchau, Ed. 1986. "Government Policies for Innovation and Growth" in National Research Council. *The Positive Sum Strategy: Harnessing Technology for Economic Growth.* Washington, D.C.: National Academy Press.